Alternative Fuels
for Road Vehicles

Alternative Fuels for Road Vehicles

M.L. Poulton

Computational Mechanics Publications
Southampton, UK and Boston, USA

M.L. Poulton
Transport Research Laboratory
Environment Centre
Old Wokingham Road
Crowthorne
Berkshire RG11 6AU

Published by

Computational Mechanics Publications
Ashurst Lodge, Ashurst, Southampton, SO40 7AA, UK
Tel: 44 (0)1703 293223 Fax: 44 (0)1703 292853
Email: CMI@uk.ac.rl.ib.
Intl Email: CMI@ib.rl.ac.uk.

For USA, Canada, and Mexico:

Computational Mechanics Inc
25 Bridge Street, Billerica, MA 01821, USA
Tel: 508 667 5841 Fax: 508 667 7582
Email: CMINA@netcom.com.

British Library Cataloguing-in-Publication Data

A Catalogue record for this book is available
from the British Library

ISBN 1-85312-301-3 Computational Mechanics Publications, Southampton
ISBN 1-56252-225-6 Computational Mechanics Publications, Boston

Library of Congress Catalog Card Number 94-70407

Printed and bound in Great Britain by Hartnolls Ltd. of Bodmin, Cornwall

PREFACE

This book has been published as a consequence of a programme of study, commissioned by the Chief Mechanical Engineer's Office at the UK Department of Transport, into the contribution of the road vehicle to global warming. A programme of research was placed with the Environment Centre of the Transport Research Laboratory, and one of the individual projects was to investigate, review and assess the future prospects for conventional and alternative fuels for road vehicle applications. Implications of the energy and emissions from the whole fuel cycle (from production to distribution and final usage) were considered but, more importantly, the vehicular fuel consumption (and consequent carbon dioxide emissions) and exhaust emission characteristics were the primary focus of attention.

The structure of this book is such that each chapter, describing a particular alternative fuel, is completely self-contained. The reader will be able to read about a particular subject that is of interest without having to refer to other chapters to gain a full understanding of the fuel's characteristics and notable developments and demonstration programmes underway worldwide. One chapter (chapter 10) does provide an overview and inter-comparison of *all* the fuels discussed, including point-of-use and life cycle emissions, global warming impacts, fuel storage implications and likely costs.

The many alternative fuels that have been reviewed are likely to be of great interest to many readers, not just mechanical, petrochemical and transportation engineers, but anyone with a technical interest and/or association with the subject – the fuel for the motor vehicle and how it may develop and change in the future. In addition to discussing the prospects for conventional petrol and diesel fuels, including their reformulation, synthetic fuels, vegetable oils and other biofuels, alcohols, gases (LPG, natural gas and hydrogen) and electricity are also considered for their future potential.

Future advances in conventional engines and the development of alternative power units are discussed in the companion volume to this book, *Alternative Engines for Road Vehicles*. The future prospects for a range of engines, including conventional petrol and diesel-fuelled units (covering technologies such as two-stroke, lean burn and stratified charge), the rotary engine, gas turbine, Stirling, Rankine (steam engine) and hybrids are assessed for their potential to reduce vehicle emissions and improve fuel economy. Other, less well known concepts such as catalytic combustion, the Quadratic (beam) engine, stepped piston and other engine thermal and mechanical efficiency-improving techniques are also reviewed.

ACKNOWLEDGEMENTS

The author's thanks are extended to the Chief Mechanical Engineer's Office at the UK Department of Transport, for its permission to use and publish much of the results of the review of alternative engines in this format. Thanks go also to the publishers and copyright holders of the illustrations used in this book.

My thanks go to colleagues in the Environment Centre at the Transport Research Laboratory (TRL) for their help and support in the writing and preparation of the text for this book. I should especially like to thank John Hickman, who read and reviewed the first draft, and Dr Kit Mitchell, who acted as Quality Audit and Review Officer for the original Department of Transport project report. Their advice and feedback on the style and contents have enabled me to make changes and include additional information to enhance the contents which now form this volume.

The efforts of Dr Phil Bly, TRL Research Director and Chris Webb from TRL Press and Public Relations, are gratefully acknowledged. They helped to enable the publication of this book and thereby make more widely available the results of a programme of scientific research and review that has been undertaken at TRL.

Mark Poulton
TRL, Crowthorne
April 1994

CONTENTS

INTRODUCTION   1

1. FOSSIL FUEL RESERVES   3
    1.1 Introduction   3
    1.2 Oil reserves   4
    1.3 Natural gas reserves   5
    1.4 Coal reserves   6
    1.5 Conclusions   7
    REFERENCES FOR CHAPTER 1   8

2. LIQUID HYDROCARBON FUELS   9
    2.1 Introduction   9
    2.2 Reformulated petrol   9
        2.2.1 Introduction   9
        2.2.2 Oxygenates   11
        2.2.3 Reformulated petrol (gasoline) research programmes   13
            2.2.3.1 US Auto/Oil Air Quality Improvement
                      Research Program (AQIRP)   13
                      2.2.3.1.1   US AQIRP Phase I summary   15
            2.2.3.2 Phillips Petroleum   16
            2.2.3.3 European Auto/Oil Research Programme   18
        2.2.4 US reformulated gasoline program   18
    2.3 Reformulated and low–sulphur diesel   18
        2.3.1 Introduction   18
        2.3.2 Important diesel fuel properties   19
            2.3.2.1 Sulphur   19
            2.3.2.2 Cetane number   20
            2.3.2.3 Density   20
    2.4 Commercially available reformulated fuels   21
    2.5 Synthetic liquid fuels   21
    2.6 The future for liquid hydrocarbon fuels   24
    2.7 Summary   25
    REFERENCES FOR CHAPTER 2   26

3. METHANOL   29
    3.1 Introduction   29
    3.2 Fuel characteristics   29
        3.2.1 Safety implications   30
    3.3 Feedstocks   31
    3.4 Infrastructure   32
    3.5 Vehicle modifications   32
        3.5.1 Petrol–engined vehicles   32
        3.5.2 Diesel–engined vehicles   34

3.6 Emissions performance 36
    3.6.1 Vehicle exhaust emissions 36
        3.6.1.1 Substituted for petrol 36
        3.6.1.2 Substituted for diesel 38
        3.6.1.3 Air toxics and secondary pollutants 40
    3.6.2 Life cycle emissions 41
3.7 Costs 42
    3.7.1 Fuel production and distribution 42
    3.7.2 Vehicle modification 42
3.8 Demonstration 43
3.9 Outlook 45
3.10 Summary 46
REFERENCES FOR CHAPTER 3 48

4. ETHANOL 53
4.1 Introduction 53
4.2 Fuel characteristics 53
    4.2.1 Safety implications 54
4.3 Feedstocks 54
4.4 Infrastructure 55
4.5 Vehicle modifications 56
    4.5.1 Petrol–engined vehicles 56
    4.5.2 Diesel–engined vehicles 57
4.6 Emissions performance 57
    4.6.1 Vehicle exhaust emissions 57
        4.6.1.1 Substituted for petrol 57
        4.6.1.2 Substituted for diesel 58
        4.6.1.3 Air toxics and secondary pollutants 59
    4.6.2 Life cycle emissions 59
4.7 Costs 60
    4.7.1 Fuel production and distribution 60
    4.7.2 Vehicle modification 61
4.8 Demonstration 62
4.9 Outlook 62
4.10 Summary 63
REFERENCES FOR CHAPTER 4 64

5. BIODIESEL AND VEGETABLE OILS 67
5.1 Introduction 67
5.2 Fuel characteristics 67
    5.2.1 Safety implications 69
5.3 Feedstocks 69
5.4 Infrastructure 70
5.5 Vehicle modifications 70
5.6 Emissions performance 71
    5.6.1 Vehicle exhaust emissions 71

		5.6.1.1	Methyl esters	71
		5.6.1.2	Unmodified vegetable oils	74
	5.6.2	Life cycle emissions		74
5.7	Costs			75
	5.7.1	Fuel production and distribution		75
	5.7.2	Vehicle modification		76
5.8	Demonstration			76
5.9	Outlook			77
5.10	Summary			78
	REFERENCES FOR CHAPTER 5			80

6. LIQUEFIED PETROLEUM GAS — 83

6.1	Introduction			83
6.2	Fuel characteristics			83
	6.2.1	Safety implications		85
6.3	Feedstocks			85
6.4	Infrastructure			86
6.5	Vehicle modifications			86
	6.5.1	Petrol-engined vehicles		86
	6.5.2	Diesel-engined vehicles		88
6.6	Emissions performance			88
	6.6.1	Vehicle exhaust emissions		88
		6.6.1.1	Substituted for petrol	88
		6.6.1.2	Substituted for diesel	91
		6.6.1.3	Air toxics and secondary pollutants	91
	6.6.2	Life cycle emissions		92
6.7	Costs			92
	6.7.1	Fuel production and distribution		92
	6.7.2	Vehicle modification		93
6.8	Demonstration			93
6.9	Outlook			94
6.10	Summary			95
	REFERENCES FOR CHAPTER 6			97

7. NATURAL GAS — 99

7.1	Introduction			99
7.2	Fuel characteristics			99
	7.2.1	Safety implications		101
7.3	Feedstocks			101
7.4	Infrastructure			101
7.5	Vehicle modifications			102
	7.5.1	Petrol-engined vehicles		103
	7.5.2	Diesel-engined vehicles		104
7.6	Emissions performance			105
	7.6.1	Vehicle exhaust emissions		105
		7.6.1.1	Substituted for petrol	105

		7.6.1.2	Substituted for diesel	108
		7.6.1.3	Air toxics and secondary pollutants	110
	7.6.2	Life cycle emissions		110
7.7	Costs			111
	7.7.1	Fuel production and distribution		111
	7.7.2	Vehicle modification		112
7.8	Demonstration			113
7.9	Outlook			115
7.10	Summary			117
	REFERENCES FOR CHAPTER 7			118

8. ELECTRICITY 123

8.1	Introduction			123
8.2	Energy storage			123
	8.2.1	Battery technology		124
	8.2.2	Lead-acid		126
	8.2.3	Nickel-cadmium		126
	8.2.4	Nickel-iron		127
	8.2.5	Sodium-sulphur		127
	8.2.6	Sodium-nickel chloride		128
	8.2.7	Lithium aluminium-iron sulphide		128
	8.2.8	Lithium aluminium-iron disulphide		128
	8.2.9	Lithium-solid polymer		129
	8.2.10	Zinc-air, aluminium-air and iron-air		129
	8.2.11	Zinc-bromine		130
	8.2.12	Nickel-metal hydride		130
8.3	Electric vehicle technology			130
	8.3.1	Electric and hybrid vehicle specifications		130
	8.3.2	Electric vehicle developments and demonstration projects		135
8.4	Electricity generation			138
	8.4.1	UK power generation		138
	8.4.2	Fuel cells		139
		8.4.2.1	Introduction	139
		8.4.2.2	Fuel cell construction and types	139
			8.4.2.2.1 Phosphoric acid fuel cell	140
			8.4.2.2.2 Proton exchange membrane fuel cell	140
			8.4.2.2.3 Solid oxide fuel cell	142
		8.4.2.3	Future fuel cell development	142
	8.4.3	Solar cells		143
8.5	Electric vehicle energy consumption and emissions			144
	8.5.1	Energy consumption		144
	8.5.2	Emissions		147
8.6	Outlook			149
8.7	Summary			150
	REFERENCES FOR CHAPTER 8			151

9. **HYDROGEN**   157
   9.1   Introduction   157
   9.2   Fuel characteristics   157
      9.2.1   Safety implications   158
   9.3   Feedstocks   158
   9.4   Infrastructure   159
   9.5   Vehicle modifications   159
   9.6   Emissions performance   161
      9.6.1   Vehicle exhaust emissions   161
      9.6.2   Life cycle emissions   162
   9.7   Costs   162
      9.7.1   Fuel production and distribution   162
      9.7.2   Vehicle modification   163
   9.8   Demonstration   163
   9.9   Outlook   164
   9.10   Summary   165
   REFERENCES FOR CHAPTER 9   166

10. **COMPARATIVE EVALUATION OF ALTERNATIVE FUELS**   169
   10.1   Introduction   169
   10.2   Energy-specific CO_2 emissions (at point-of-use)   169
   10.3   Fuel cycle CO_2 and greenhouse gas emissions   171
      10.3.1 Greenhouse gas emissions and their effects   171
      10.3.2 Alternative fuels - fuel cycle greenhouse
          gas emissions   173
      10.3.3 Alternative fuels - emissions from production and
          distribution as a proportion of total greenhouse gases   177
   10.4   Exhaust (pollutant) emissions from alternative fuels   178
   10.5   Vehicle storage implications of alternative fuels   180
   10.6   Costs of alternative fuels   182
   REFERENCES FOR CHAPTER 10   184

11. **HEALTH AND ENVIRONMENTAL EFFECTS OF**
EXHAUST EMISSIONS   187
   11.1   Introduction   187
   11.2   Hydrocarbons   187
   11.3   Non-methane hydrocarbons   187
   11.4   Non-methane organic gases   187
   11.5   Volatile organic compounds   188
   11.6   Methane, ethane and propane   189
   11.7   Methanol and ethanol   190
   11.8   Aldehydes   190
   11.9   Olefins   192
   11.10 Aromatics   192
   11.11 Nitrogen oxides   194
   11.12 Carbon monoxide   195

11.13 Particulates   196
11.14 Lead   198
11.15 Ozone and PAN   199
11.16 Carbon dioxide   200
11.17 IARC categories   200
REFERENCES FOR CHAPTER 11   202

12. SUMMARIES AND CONCLUSIONS   205
12.1 Introduction   205
12.2 Liquid hydrocarbons   205
12.3 Methanol   206
12.4 Ethanol   207
12.5 Biodiesel and vegetable oils   208
12.6 Liquefied petroleum gas   208
12.7 Natural gas   208
12.8 Electricity   209
12.9 Hydrogen   210
12.10 Conclusions   211

INDEX   213

LIST OF FIGURES

1A	Proved reserves: oil	3
1B	Proved reserves: natural gas	3
2A	Reserves/production ratios: oil	3
2B	Reserves/production ratios: natural gas	3
3	Effect of AQIRP petrol reformulation on exhaust mass emissions	13
4	Effect of AQIRP petrol reformulation on exhaust toxics mass emissions	13
5	Effect of 1 psi RVP reduction on exhaust and evaporative mass emissions	14
6	Effect of oxygenates on exhaust and evaporative mass emissions	14
7	Effect of oxygenates on exhaust toxics mass emissions	15
8	AGO blending components	20
9	Schematic diagram of VW multi-fuel concept	33
10	Effects of using M85 in prototype FFVs/VFVs	37
11	DDC methanol bus engine	44
12	Exhaust emissions results from a RME-fuelled 2.3-litre diesel engine	73
13	A bi-fuel LPG system with three-way catalyst	87
14	The influence of air/fuel ratio on exhaust emissions with LPG fuel	89
15	Emissions from LPG-fuelled Opel Omega in the ECE cycle	90
16	Layout of CNG components in a petrol van conversion	104
17	Predicted UK electricity consumption and generation mix 1990–2005	138
18	Efficiency of 30 kW internal combustion engine and PEMFC with reformer	141

19 EV energy consumption as a function of vehicle mass   145

20 EV energy consumption as a function of distance travelled   146

21 Point–of–use CO_2 emissions from the combustion of
    alternative fuels for optimised engine designs   170

22 The relative greenhouse contributions from all
    man–made emissions   173

23 Estimates of ranges of greenhouse gas emissions (CO_2 equivalent)
    from the fuel cycle analysis of alternative fuels   174

24 Alternative fuels – fuel production and distribution contribution
    to total greenhouse gas emissions   177

25 Vehicle on–board storage implications of alternative fuels   181

26 Estimates of the ranges of overall alternative fuel costs   183

LIST OF TABLES

1 Proved conventional oil reserves, end 1991   4

2 Proved natural gas reserves, end 1991   5

3 Proved coal reserves, end 1991   6

4 Blending characteristics of fuel oxygenates   11

5 EC oxygenate blending limits   12

6 Results of Phillips Petroleum reformulated gasoline programme: percentage change from conventional fuel to *Unleaded Plus*   16

7 DDC 6V–92TA methanol engine certified exhaust emissions   39

8 DDC 6V–92TA ethanol–fuelled engine certified exhaust emissions   59

9 Exhaust emission results from US EPA heavy–duty transient testing of rapeseed and soybean methyl esters   71

10 Exhaust emission results from IFP and UTAC testing of heavy and light–duty diesel engines fuelled with RME   72

11 Volvo 940 passenger car closed–loop LPG emissions performance   90

12 LPG–fuelled heavy–duty engine exhaust emissions   91

13 Lean–burn natural gas–fuelled passenger car emissions performance   107

14 Three–way catalyst natural gas–fuelled passenger car emissions performance   107

15 Cummins L10 lean–burn natural gas engine emissions performance   109

16 Iveco 8469.21 three–way catalyst–equipped natural gas engine emissions performance   110

17 USABC primary criteria for EV battery performance   124

18 EV battery technology status   125

19 Electric and hybrid vehicle specifications   131–135

20 Comparison of electric vehicle energy consumption figures   144

21 SAE J227a recommended electric vehicle test schedules   146

22 UK power generation greenhouse gas emissions:
    g per kWh delivered   148

23 Possible EV emissions performance for UK power supply mix   148

24 Implications of hydrogen storage relative to petrol   161

25 Specific heat output and carbon dioxide emissions at
    point of use for various fuels   170

26 IPCC estimates of Global Warming Potentials for various gases   172

27 Pollutant emissions performance of alternative fuels   179

28 Contribution of road transport sources to VOC emissions in the UK   188

29 IARC carcinogenicity categories   201

ALTERNATIVE FUELS FOR ROAD VEHICLES

INTRODUCTION

Liquid hydrocarbon fuels are very well suited to road transport applications. They are easily handled and possess a high energy density. As a consequence, conventional petroleum and diesel fuels have remained almost entirely unchallenged since the motor vehicle was invented. However, their source is finite, reserves are not uniformly distributed and their increased usage contributes to a variety of local and regional air pollution problems and potential climate change.

Several alternative forms of fuel have been used in the motor vehicle since the production of petrol and diesel fuels first began. Most offer the prospect of reduced levels of certain combustion-related exhaust emissions and other benefits may include better fuel consumption (reduced emissions of carbon dioxide, CO_2) and ensuring diversity of fuel supply by employing non-conventional feedstocks. Some alternatives use other carbon-based fuels (for example, natural gas) either as a fuel directly or convert it to other forms, such as methanol.

Very few alternative fuels have been exploited commercially on a large scale. Two main problems have been the enormous infrastructure requirements for the production, distribution and refuelling aspects of using an alternative, plus the fact that cost has been a major barrier when compared with conventional liquid hydrocarbon fuels. Certain fuels (electricity and natural gas, for example) have production facilities and distribution networks already in place in most developed countries, and are more likely to overcome market barriers. Nevertheless, additional supply and distribution capacity and certain additional refuelling components would be needed if these energy supplies were used as a vehicular fuel on a large scale. Other liquid alternatives pose a range of problems in terms of their production and distribution, although liquid fuels may lend themselves more easily to a new or modified existing distribution network than gaseous ones.

Internationally, there is a considerable quantity of information on research and practical experience of the production and vehicular use of alternative fuels. The purpose of this book is to draw on existing information and review the various alternative fuels that exist for road vehicles.

For each alternative considered, benefits and disadvantages are discussed with respect to exhaust emissions, energy consumption, fuel production and distribution implications and vehicular modifications that may be necessary. The book focuses on the vehicular aspects of emissions (regulated, CO_2 and others); the complete fuel cycle aspects are highlighted only, since they are not the subject of the study upon which this book is based. Cost issues are made generally, with definitive data where it exists.

1. FOSSIL FUEL RESERVES

1.1 Introduction

This chapter presents an overview of the scale and geographical location of proven fossil fuel (oil, natural gas and coal) reserves. The tables included later in this chapter present only a "snapshot" view (as at the end of 1989 and 1991) of fossil fuel reserves and their expected life at current production/ consumption rates. Historical data for the quantities of proved reserves of oil and natural gas are shown in Figure 1 and the reserves/ production ratio (years of reserves remaining) in Figure 2 (both from British Petroleum, 1992).

Over the last 25 years the reserves to production ratio for oil has fluctuated, but has averaged nearly 35 years, indicating that new reserves have continually been discovered

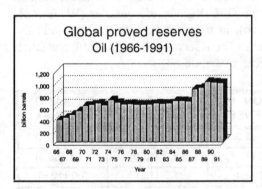

Figure 1A. Proved reserves: oil

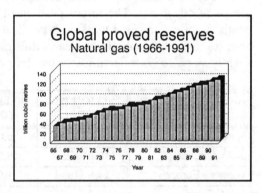

Figure 1B. Proved reserves: natural gas

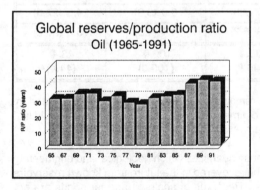

Figure 2A. Reserves/production ratios: oil

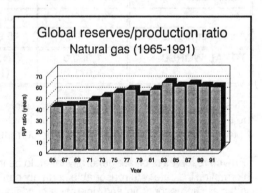

Figure 2B. Reserves/production ratios: natural gas

in line with, and sometimes at a faster rate than current production/consumption, thus increasing the ratio. Over the last few decades the reserves to production ratio for natural gas has also fluctuated, but steadily upward, although it appears to be levelling off.

The *current* (end of 1991) geographical distribution of proved oil, natural gas and coal reserves are shown in Tables 1, 2 and 3. At the current rate of consumption of oil and natural gas the remaining known reserves would last for about another 40 and 60 years respectively. Coal, however, mined at the current rate, would last some 240 years.

1.2   Oil reserves

Table 1 (from British Petroleum, 1992) gives the current known reserves of oil for various groups of countries. The minor position of the OECD group of countries (W. Europe, North America, Japan, Australia, New Zealand), in terms of both oil reserves and the reserves to production ratio, is clearly illustrated. The reserves/production ratio has fallen slightly since 1989, reflecting low market prices for crude oil.

AREA	BILLION (10^9) BARRELS	BILLION (10^9) TONNES	SHARE OF TOTAL (%)	R/P RATIO[1] YEARS
North America	41.7 (42.4)	5.3 (5.3)	4.2 (4.2)	10.2 (10.4)
Latin America	119.8 (125.2)	16.9 (17.6)	12.0 (12.5)	43.1 (50.9)
Western Europe	14.5 (18.4)	1.9 (2.4)	1.5 (1.8)	9.0 (12.6)
CIS/E Europe	58.8 (59.9)	8.0 (8.2)	5.8 (5.9)	15.1 (13.0)
Middle East	661.6 (660.3)	89.4 (89.3)	66.1 (65.2)	100+ (100+)
Africa	60.4 (58.8)	8.0 (7.8)	6.0 (5.9)	24.5 (27.5)
Asia/Australasia	44.1 (46.8)	5.9 (6.2)	4.4 (4.5)	18.3 (20.2)
TOTAL WORLD	**1,000.9 (1,011.8)**	**135.4 (136.8)**	**100.0 (100.0)**	**43.4 (44.4)**

[1] Reserves/Production (R/P) Ratio: if the reserves remaining at the end of any year are divided by the production in that year, the result is the length of time that those remaining reserves would last if production were to continue at the then current level.

[2] Proved reserves of oil are generally taken to be those quantities which geological and engineering information indicate with reasonable certainty can be recovered in the future from known reservoirs under existing economic and operating conditions. Reserves of shale oil and tar sands are not included.

Table 1:   Proved[2] conventional oil reserves, end 1991 (end 1989)

It is noteworthy that world proven oil reserves five years earlier were about 700 billion barrels (over 300 billion barrels less than current estimates) with a world reserves to production ratio in 1986 of 32.5 years. Substantial increases in proven reserves have occurred between 1986 and 1989, mainly due to upward revisions by Iran, Iraq, Abu Dhabi and Venezuela (IEA, 1990).

1.3 Natural gas reserves

Natural gas is likely to play an important role if there is to be diversification of transport fuels in the future. Natural gas can be used as compressed natural gas (CNG) or liquefied natural gas (LNG) or alternatively converted into methanol, petroleum or distillates. It is often seen as a more "abundant" source of supply than oil as Table 2 (from British Petroleum, 1992) illustrates, although only in terms of the reserves/production ratio at the *current rate* of production/consumption. The actual quantity (in oil equivalent) is slightly less than crude oil; natural gas is not consumed so quickly at present.

AREA	TRILLION (10^{12}) CUBIC METRES	BILLION (10^9) TONNES OIL EQUIV[1]	SHARE OF TOTAL (%)	R/P RATIO[2] YEARS
North America	7.5 (7.4)	6.7 (6.6)	6.1 (6.5)	12.3 (12.6)
Latin America	6.8 (6.7)	6.1 (6.0)	5.4 (5.8)	69.2 (75.0)
Western Europe	5.1 (5.4)	4.6 (4.9)	4.1 (4.9)	25.7 (31.3)
CIS/E Europe	50.0 (43.3)	45.0 (39.0)	40.4 (38.3)	58.9 (50.7)
Middle East	37.4 (34.7)	33.7 (31.3)	30.1 (30.7)	100+ (100+)
Africa	8.8 (7.5)	7.9 (6.8)	7.1 (6.7)	100+ (100+)
Asia/Australasia	8.4 (8.0)	7.6 (7.2)	6.8 (7.1)	48.2 (55.3)
TOTAL WORLD	124.0 (113.0)	111.6 (101.7)	100.0 (100.0)	58.7 (56.3)

[1] Based on 1.111 trillion (10^{12}) cubic metres of natural gas = 1 billion (10^9) tonnes oil equivalent.

[2] Reserves/Production (R/P) Ratio: if the reserves remaining at the end of any year are divided by the production in that year, the result is the length of time that those remaining reserves would last if production were to continue at the then current level.

[3] Proved reserves of natural gas are generally taken to be those quantities which geological and engineering information indicate with reasonable certainty can be recovered in the future from known reservoirs under existing economic and operating conditions.

Table 2: Proved[3] natural gas reserves, end 1991 (end 1989)

OECD countries possess a much greater proportion of proved world gas reserves than oil. In addition, the number of individual sources of supply is greater for gas than for oil and their distribution is different, giving consumers greater choice and hence reducing the possibility of control by cartels (IEA, 1990). The OECD reserves/production ratio for gas is almost double that for oil, despite the fact that exploration for gas has historically been much less than for oil; this is largely because natural gas is often found when the prime target is oil. As exploration is costly, companies do not generally explore until they perceive a future market; this tends to limit the reserves to production ratios.

Natural gas is a major energy source with substantial known reserves, which could in theory supply directly, or indirectly as a feedstock for other fuels, a significant part of road transport demand. This contrasts with some other alternative fuels where their availability may restrict them to limited application.

1.4   Coal reserves

Table 3 (from British Petroleum, 1992) shows the known global reserves of coal and their oil equivalent. The greatest contrast between oil (and to a lesser extent natural gas) and coal reserves is the geographical distribution, and is therefore less likely to be a constraint from an energy security point of view. The scale of reserves indicates that based on oil equivalent, there is almost four times the energy stored in global coal reserves than in oil.

AREA	BILLION (10⁹) TONNES		BILLION TONNES OIL EQUIV[1]	SHARE OF TOTAL (%)	R/P RATIO[2] YEARS
	Anthracite & Bituminous	Sub–bituminous & Lignite			
North America	117.2	132.0	122.1	23.9	260
Latin America	6.9	4.5	6.1	1.1	278
Western Europe	29.3	69.2	42.6	9.5	166
CIS/E Europe	136.2	179.3	150.6	30.4	306
Middle East	0.2	0.0	0.1	<0.05	N/A
Africa	60.8	1.3	41.0	6.0	336
Asia/Australasia	170.8	132.8	158.1	29.1	196
TOTAL WORLD	521.4	519.1	520.6	100.0	239

[1] Based on 1.5 million (10⁶) tonnes of coal = 1 million (10⁶) tonnes oil equivalent and 3.0 million (10⁶) tonnes of lignite = 1 million (10⁶) tonnes oil equivalent.

[2] Reserves/Production (R/P) Ratio: if the coal reserves remaining at the end of any year are divided by the production in that year, the result is the length of time that those remaining coal reserves would last if production were to continue at the then current level.

[3] Proved reserves of coal are generally taken to be those quantities which geological and engineering information indicate with reasonable certainty can be recovered in the future from known deposits under existing economic and operating conditions.

Table 3.   Proved[3] coal reserves, end 1991

Coal can be used as a fuel itself, or as a feedstock for the production of other liquid fuels, such as synthetic petroleum or methanol. The conversion of coal to methanol is a well established process but, as chapter 10 demonstrates, is not as energy efficient as that using other feedstocks such as natural gas.

1.5   Conclusions

This chapter has been included to present an overview of the magnitude and geographical distribution of proven fossil fuel reserves. It is useful to appreciate the scale of the effect that a switch to natural gas (or other alternative fuels using methane as a feedstock), for example, would have on the reserves/production ratio (which would fall), with consequent benefits to crude oil supplies. It is also useful to realise the scale of the contribution that biomass and renewable fuel–switching may have on extending fossil fuel supplies – bearing in mind that fossil fuels are being consumed by all sectors (based on 1991 consumption) at a rate of **7.2 billion tonnes oil equivalent** (toe) per annum worldwide.

While this book does not give examples of the effects of fuel switching and substitution, the reader will be able to make estimates of the consequences from the data presented in this chapter and from statistics showing road transport's share of total fossil fuel consumption. This book does not attempt to estimate the degree to which road transport might become a premium user of fossil fuels, particularly oil.

REFERENCES FOR CHAPTER 1

BRITISH PETROLEUM (1992). BP Statistical Review of World Energy. British Petroleum Company plc, Corporate Communications Services, London.

INTERNATIONAL ENERGY AGENCY (1990). Substitute fuels for road transport. A technology assessment. OECD/IEA, Paris.

2.   LIQUID HYDROCARBON FUELS

2.1   Introduction

Liquid hydrocarbon fuels are very well suited to road transport applications and conventional petroleum and diesel fuels have remained almost entirely unchallenged since the motor vehicle was invented. Their continued and increased use, however, contributes to a variety of local and regional air pollution problems and potential global climate change. The use of alternative fuels may help to improve local air quality, reduce regional pollution (such as "acid rain"), mitigate or reduce possible global warming and reduce the dependence on finite supplies of fossil fuels. Consequently, substantial research and development into alternatives is taking place.

Nonetheless, crude oil–based fuel products are very well established with the infrastructure in place to fuel virtually every motor vehicle operating today. The oil companies are expected, therefore, to also put considerable development efforts into reducing the environmental impact from the use of their products, in preference to witnessing legislation to force the introduction of alternatives fuels.

This chapter discusses conventional liquid hydrocarbon fuels – petroleum and diesel fuels as we know them today, and considers what changes and resulting benefits may take place in the future. Topics covered include reformulated petrol, including the addition of oxygenates such as MTBE or ETBE, and low–sulphur (and reformulated) diesel fuels. Also, the prospects and implications for conventional fuel substitution by synthetic liquid hydrocarbon fuel (manufactured from coal or shale oil, for example) are discussed.

2.2   Reformulated petrol

2.2.1   Introduction

Petrol is a complex mixture of flammable liquid hydrocarbons. Some of these hydrocarbons are present in crude oil and are obtained using distillation and other separation technologies. Others are created by a variety of physical and chemical transformation processes, often in the presence of catalysts, in a modern refinery. For example, aromatics are obtained from catalytic reforming and olefins from catalytic cracking or catalytic polymerisation. For use in a modern petrol engine, petrol must satisfy a number of requirements regarding its volatility, octane level and other characteristics. Refiners can satisfy these requirements using a wide variety of different combinations of chemical constituents, with their selection dependent on relative costs of the different components, market prices for other products, refinery capability and the quality of crude oil available.

The production of reformulated petrol accentuates the importance of a particular fuel characteristic - the fuel's effect on emissions, in the selection of petrol components. For example, the addition of oxygenates (see section **2.2.2**) to the petrol blend can reduce carbon monoxide emissions and may serve to reduce the exhaust hydrocarbon (HC) reactivity. Reducing the more volatile components of the fuel will reduce overall volatility and result in lower evaporative emissions. The four major constituent groups of petrol are:

> • **olefins** - high octane components produced from crude during refining and also occurring naturally in low concentrations in crude. Many olefins are both highly volatile and highly reactive;

> • **aromatics** - higher octane (than olefins) constituents, occurring naturally in crude in moderate to high concentrations and also created by refining. Aromatics are reactive, though less so than olefins;

> • **paraffins** - consisting of two groups, *highly branched* paraffins that are both high in octane and low in reactivity and *straight chain* paraffins that are also low in reactivity but are very low octane. Paraffins, like aromatics, are present in crude in moderate to high concentrations, depending on crude type; and

> • **napthenes** - between paraffins and olefins in octane and present in crude in moderate to high concentrations.

Petrol is reformulated to provide unleaded fuel, since removing lead creates a need for additional octane enhancement. Refiners have increased the conversion of paraffins and napthenes into higher octane branched paraffins, olefins and aromatics. Ironically, these changes, designed to allow the use of catalytic converters, increase both the reactivity and volatility of the fuel.

In August 1989 the Atlantic Richfield Oil Company (ARCO) introduced a so-called "reformulated gasoline" to replace regular leaded petrol in southern California. Called EC-1, it contains one-third less olefins and aromatics and 50 percent less benzene, no lead and 80 percent less sulphur than the regular petrol. The vapour pressure is 1 psi lower than the US South Coast standard and it contains the oxygenate MTBE. ARCO has claimed significant emissions reductions when EC-1 is used in place of regular leaded petrol in pre-1975 cars (without catalytic converters), with evaporative emissions 21 percent lower, CO down nine percent, NO_x five percent lower and four percent less HC emissions (US Congress, 1990).

ARCO redirects the olefin and aromatic streams removed from EC-1 into its unleaded grades used by cars with catalytic converters, which are thought to emit relatively small additional pollution compared with the amount saved by the uncontrolled cars. However, as the fleet turnover progresses, the numbers of uncontrolled vehicles will dwindle to virtually nil and the relatively large total emissions savings will fall. For this reason, many observers and petroleum companies believe that reformulated fuels should only be produced to fill niches (such as older cars) as and when desirable, and not be forcibly introduced by legislation (Cragg, 1992).

It has been generally accepted that reformulation of petrol for environmental reasons lowers volatility, lowers the concentrations of aromatics (especially benzene, due to its carcinogenicity) and volatile olefins, and adds oxygenates. However, the precise role that each petrol constituent plays in vehicle emissions is not well known. Catalytic emission controls further complicate the relationship between petrol composition and emissions by destroying some HCs and converting others into new compounds with different reactivities. While the desirability for reduced aromatic and olefin content is accepted, refiners have been unable to quantify the effects of reductions, especially as aromatics and olefins are produced during combustion and in the catalyst.

To better understand the impact of changes in the major petrol constituents on emissions, three US automotive companies and 14 petroleum companies began a joint research programme – the Auto/Oil Air Quality Improvement Research Program (AQIRP); this major research project and its findings are described in section **2.2.3.1**.

2.2.2 Oxygenates

Oxygenates, being alcohols or ethers, are generally so–called because of their high molecular oxygen content. The alcohol oxygenates are methanol, ethanol and tertiary butyl alcohol (TBA) and the ethers are methyl tertiary butyl ether (MTBE), ethyl tertiary butyl ether (ETBE) and tertiary–amyl methyl ether (TAME). They are not normally the product of crude oil but are most commonly produced from natural gas or distilled from biomass. Table 4 lists the blending characteristics of fuel oxygenates, including minimum oxygen content set by the US Environmental Protection Agency (EPA) for phase I reformulated and oxygenated petrol (from Cragg, 1992).

	Vapour pressure psi	Octane rating (RON+MON)/2	Oxygen content % wt.	US EPA blending limits	
				% vol.	Oxygen % wt.
Methanol	60	120	50	–	–
Ethanol	19	115	35	10	3.7
MTBE	8	110	18	15	2.7
ETBE	4	111	16	13	2.0

Table 4. Blending characteristics of fuel oxygenates

Oxygenates have been quoted as an important element in the emissions reduction potential of future petrol. Since oxygenates contain a considerable quantity of oxygen, their addition to petrol should ensure better and more complete fuel combustion leading especially to lower emissions of CO and HCs. In the US, several cities require petrol (gasoline) to contain oxygenates corresponding to two percent oxygen by weight (US Congress, 1990). The effects of oxygenates on exhaust emissions are shown in Figure 6 in section **2.2.3.1,** as demonstrated by Stage I of the US Auto/Oil Air Quality Improvement Research Program.

Oxygenates also have advantages in higher octane ratings than petrol – the table lists the more realistic measure of *road octane*, which is an average of RON and MON (a similar anti–knock test but performed at higher engine speed and temperature and invariably lower). Consequently a road octane number of 110 represents around RON 115–120 (Cragg, 1992). The use of oxygenates as high octane blending components (in lieu partly of aromatics and olefins) is increasing, as demonstrated in the USA. Tests with prototype oxygenated fuels in European cars have shown that equal or superior anti–knock performance was obtained with ethers, and less satisfactory results from the alcohols (Williams and Vincent, 1992). However, methanol and ethanol, in low volume, are frequently blended with petrol, usually as an extender (such as to absorb surplus methanol production in Germany).

Methanol and ethanol contain high levels of oxygen and, when blended with petrol, reduce the calorific value of the fuel and thereby worsen fuel economy. The alcohols also have high vapour pressures (especially methanol) which, when blended with petrol, increase the overall volatility of the fuel and hence evaporative emissions. Two of the key oxygenates that are promoted for use in petrol are MTBE and ETBE since they have sufficiently low vapour pressures, high octane rating and a reasonable oxygen content.

EC Directive 85/536/EEC imposed permissable limits for oxygenate blends in European Community petrol, primarily to help reduce over–consumption of crude oil necessary in refining to produce lead–free petrol (OJEC, 1985). Member States must freely permit, for the proportions by volume of organic oxygenate compounds in fuel blends, those not exceeding limits as shown in the central column in Table 5.

Oxygenate	Permissable limit % vol.	Requirement for marking at the pump > % vol.
Methanol	3.0	3.0
Ethanol	5.0	5.0
Ethers (MTBE, ETBE, TAME)	10.0	15.0
Mixture of oxygenates	2.5% oxygen by weight	3.7% oxygen by weight

Table 5.   EC oxygenate blending limits

EC Member States may authorise higher blending proportions than those listed, although petrol dispensing pumps must be clearly marked to indicate the variation of the fuel's calorific value if the oxygenate levels are *above* the limits shown in the right-hand column of Table 5.

2.2.3 Reformulated petrol (gasoline) research programmes

2.2.3.1 US Auto/Oil Air Quality Improvement Research Program (AQIRP)

Phase I of the US AQIRP, established in 1989, was intended to determine, through research and testing, the potential reduction in vehicle emissions and improvements in air quality (primarily ozone), and the relative cost effectiveness from the use of reformulated gasoline and methanol as M85 in flexible fuelled vehicles (FFVs). Phase II was to examine the effects on vehicle emissions of the best reformulated petrols determined from Phase I results.

Phase I involved more than 2,200 emissions tests using 29 fuel compositions in 53 vehicles (CRC, 1993). Two fleets of vehicles were established - then current (in 1989) and older (1983-85) vehicles. The test fuels' compositions were reformulated with respect to aromatic and olefin content, oxygenate addition, sulphur content, volatility (Reid Vapour Pressure - RVP) and the 90 percent distillation temperature (T_{90}). The Stage I results for a fleet of 20 1989 model year (MY) vehicles (with emissions control equipment expected to be found on US production vehicles throughout the first half of the 1990s) are reported by Colucci and Wise (1992). Figures 3 and 4 (from Colucci and Wise, 1992) show the effect of petrol reformulation on mass emissions and on toxics.

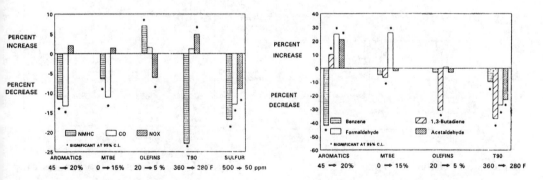

| **Figure 3.** | Effect of AQIRP petrol reformulation on exhaust mass emissions | **Figure 4.** | Effect of AQIRP petrol reformulation on exhaust toxics mass emissions |

The presence of * in Figures 3 and 4 (and also 5, 6 and 7) indicates that a significant effect on emissions was measured with 95 percent confidence.

The effects on mass emissions are caused by the reformulation indicated underneath each. Reducing aromatics and adding MTBE significantly reduced NMHC and CO emissions. Reduced olefin content increased NMHC and reduced NO_x. Although reducing olefins increased exhaust mass NMHC emissions, the ozone–forming potential of the total vehicle emissions was reduced. Reducing T_{90} had the largest impact on NMHC of all fuel variables studied, although NO_x increased slightly (Hochhauser *et al,* 1991). Additional studies were planned as part of Phase II to identify the heavy HCs that are responsible for the T_{90} effect. Sulphur was the only fuel property whose reduction caused significantly lower levels in all three exhaust emission constituents, NMHC, CO and NO_x (Benson *et al,* 1991).

Figure 4 shows that the reduction in aromatics caused the largest reduction in benzene. The greatest reduction in 1,3–butadiene came from reductions in olefins and T_{90}. Formaldehyde increased with reduced aromatics and the addition of MTBE, but decreased with the reduction in T_{90}. Acetaldehyde increased with reduced aromatics and decreased with a reduction in T_{90} also (Gorse *et al,* 1991).

The effect of reducing RVP by one psi varied, depending on the reformulation. Petrol without oxygenate and with 10 percent ethanol blend showed significantly lower diurnal (i.e. due to ambient temperature changes and hot soak caused by engine heat) emissions and lower CO without oxygenate, and lower NMHC for the ethanol blend (Reuter *et al,* 1992). These effects are shown in Figure 5 (from Colucci and Wise, 1992). No significant effect on toxic emissions was detected by the lower RVP.

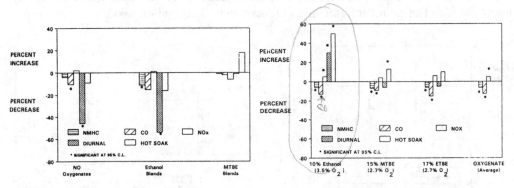

Figure 5.   Effect of 1 psi RVP   **Figure 6.**   Effect of oxygenates on
reduction on exhaust and   exhaust and evaporative
evaporative mass emissions   mass emissions

Figure 6 (from Colucci and Wise, 1992) shows the effect of adding each oxygenate (to the same base petrol) on exhaust and evaporative emissions. All reduced NMHCs and CO significantly, but increased NO_x. Figure 7 (ibid) overleaf demonstrates the effect of oxygenates on exhaust toxics mass emissions. The significant effects of ethanol and

ETBE on acetaldehyde emissions can be clearly seen. MTBE raised formaldehyde emissions but, as Figure 7 shows, by no more than ethanol or ETBE. All three oxygenates considered had the effect of lowering benzene emissions, and ethanol also lowered 1,3-butadiene.

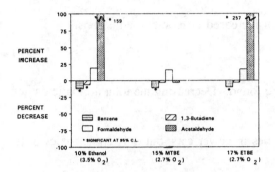

Figure 7. Effect of oxygenates on exhaust toxics mass emissions

The major effects of fuel compositional changes on mass emissions (regulated and air toxics), determined from the first phase of the US Auto/Oil Air Quality Improvement Program, are summarised below in **2.2.3.1.1** (CRC, 1993). The second phase of the US AQIRP was due for completion during 1993, and was studying in greater detail some of the effects of the important fuel compositional properties. The program was also to investigate how future production vehicles respond to changes in petrol composition. Alternative fuels were to be explored in greater depth during Phase II by analysing emissions from nominally 1993 FFV/VFV (variable fuel vehicle) fleet demonstration vehicles designed for M85 fuel, and from dedicated alcohol, LPG and compressed natural gas vehicles. At the time of writing this book, the report on the findings of Phase II had not been published.

2.2.3.1.1   US AQIRP Phase I summary

Regulated emissions:

● lowering T_{90} reduced exhaust HC by 22 percent in current vehicles and by a smaller amount in older vehicles. It raised NO_x by 5 percent in current vehicles;

● reducing the sulphur content reduced exhaust HC, CO and NO_x emissions in the current vehicle fleet (this was not tested in the older vehicles);

● decreasing RVP by 1 psi (from 9 to 8 psi) reduced evaporative emissions and also produced small reductions in exhaust HC and CO;

• decreasing olefins increased exhaust HC and reduced NO_x, and also reduced the photochemical reactivity of exhaust and evaporative emissions;

• increasing MTBE reduced exhaust HC and CO in both fleets, and raised NO_x in low aromatic fuels. Other oxygenates tested (ETBE and ethanol) produced similar results;

• raising RVP and splash blending ethanol increased evaporative emissions.

Air toxics:

• benzene was the most prevalent air toxic found. Decreasing the total aromatic content reduced benzene emissions;

• decreasing T_{90} reduced all measured toxics in newer cars, but only some toxics in the older fleet;

 • decreasing sulphur reduced exhaust toxics by 10 percent;

• oxygenate addition increased aldehyde emissions in all cars.

2.2.3.2   Phillips Petroleum

The results of another US reformulated gasoline research programme, using Phillips Petroleum *Unleaded Plus,* in vehicles spanning a range of emissions control technologies, has been reported by Schoonveld and Marshall (1991). Passenger cars, spanning four emissions control technologies, were tested using mid–grade (89 Octane) unleaded gasoline and Phillips' reformulated product (5.3% MTBE providing ≥1.0% fuel oxygen; max. 0.9% benzene; max. 20% aromatics; RVP reduced by 0.7 psi). The results are shown in Table 6 below.

Vehicle technology class	Regulated emissions			Fuel economy	Toxic emissions			
	CO	Total HC (exh+evap)	NO_x		Formald–ehyde	Acetald–ehyde	1,3 Butadiene	Total benzene
Pre–1975	**−5.4**	+1.4	**−9.3**	**−4.6**	**+20.3**	**+15.2**	–	**−35.8**
1975–1980	**−9.8**	**+13.7**	**−4.8**	**−2.7**	**+57.8**	**+19.0**	−5.5	**−25.3**
1981–1986	**−12.6**	**−14.8**	**−7.9**	−0.4	+4.4	**+32.9**	**−39.6**	**−42.4**
Post–1986	**−23.0**	**−27.9**	**−7.9**	**−3.3**	**+48.4**	−8.3	**−34.9**	**−53.3**

Note: statistically significant results @ 95% confidence level shown **bold**

Table 6.   Results of Phillips Petroleum reformulated gasoline programme: percentage change from conventional fuel to *Unleaded Plus.*

The (US) emissions control technologies included non–catalyst (pre–1975 vehicles), open loop oxidation catalyst (1975–1980 vehicles), three–way catalyst without adaptive learning (1981–1986 vehicles) and three–way catalyst with adaptive learning (post–1986 vehicles).

The Phillips Petroleum results are shown in Table 6, with the statistically significant changes in emissions (at the 95% confidence level) shown in bold type. Results for the pre–1975 and post–1986 vehicle classes were statistically significant in more cases than for the other technology classes and it was reported that emission measurements for the post–1986 vehicles showed good repeatability for each car/fuel combination. Results for changes in HC and benzene emissions in Table 6 are **total** emissions (i.e. exhaust plus evaporative).

All the vehicle technology classes showed reductions in exhaust CO and NO$_x$ with the reformulated gasoline, ranging from 5.4 percent (non–catalyst) to 23 percent in the newest technology class. Although the percentage reduction was greater for the new vehicles, the *actual* grams per mile pollutant reduction was much larger for the older cars since their emission rates were much higher. The percentage NO$_x$ reductions with the reformulated gasoline were more consistent across technology classes with the older vehicles again exhibiting the greatest grams per mile reduction. A fuel economy loss (because of the reformulated fuel's lower calorific value) of between 0.4 and 4.6 percent was reported.

The total HC emissions results were variable across the technology classes. All post–1980 vehicles showed a reduction in total HCs, as high as 28 percent for the newest technology (mostly attributable to exhaust emissions; evaporative changes were negligible). The non–catalyst vehicles showed a minimal change in HC emissions and the open–loop (oxidation) catalyst vehicles demonstrated an increase (although a vapour hose leak was discovered in one of the two vehicles in this technology class).

The US Clean Air Act (CAA) Amendments of 1990 identified five specific toxic compounds to be considered in the evaluation of reformulated fuels. They are formaldehyde, acetaldehyde, 1,3 Butadiene, benzene and polycyclic organic matter (POM). All but the POM toxics were measured in the Phillips' study. All vehicle technologies experienced increases in exhaust formaldehyde emissions when using the reformulated fuel, but all the post–1980 vehicles emitted (mostly very much) less than the maximum formaldehyde emission of 15 mg/mile specified for "Clean Fuel Vehicle Emissions" in the CAA Amendments. Acetaldehyde emissions also increased across the three oldest technology classes. The averaged 1,3 butadiene emission reductions ranged from 0 to almost 40 percent when using the reformulated fuel, with the newest vehicles displaying the largest falls. All vehicles exhibited large reductions in benzene emissions (mostly in the exhaust; evaporative changes were small).

While the Phillips Petroleum study was limited in scope (two vehicles in each technology class – 8 vehicles in total, although these were popular, high–selling models), the study's results showed that the use of reformulated fuels across a range of technologies can produce emission reductions in both their mass and reactivity (ozone forming potential) while reducing (certain) air toxics when compared with conventional fuels.

2.2.3.3   European Auto/Oil Research Programme

The European Automobile Manufacturers Association (ACEA) and the European Petroleum Industry Association (EUROPIA) reached agreement in July 1993 to establish a two–year research programme to investigate the effects of fuel specifications and engine technologies on vehicle exhaust emissions (ACEA/EUROPIA, 1993).

The agreement grew from concern about improving the environmental performance of the vehicle manufacturers' and petroleum industry's products, in addition to providing scientific evidence to be used in the setting of EC emission standards (for vehicles and fuels) for the year 2000. The programme's aims are to identify relationships between fuel properties, engine technologies, vehicle emissions, fuel consumption and life cycle CO_2 emissions, and to identify the relative importance of vehicle emissions on air quality at urban and regional level.

2.2.4 US reformulated gasoline program

The United States Clean Air Act of 1990 mandated changes in the composition of gasoline sold in the US to meet the requirements of two separate pollution reduction programmes (Petroleum Review, 1993). The oxygenated fuels programme began in November 1992 and requires gasoline sold during the winter months to contain at least 2.7 percent by weight of oxygen (in the form of blended oxygenates) in the (40) major US cities classified as carbon monoxide non–attainment areas.

The reformulated gasoline program is scheduled to commence in January 1995 in nine metropolitan areas with high ozone levels, and will affect fuel composition by imposing performance standards and specifying the content of benzene, aromatics, oxygen, heavy components and detergents, for year–round use.

2.3   Reformulated and low–sulphur diesel

2.3.1 Introduction

Diesel oil, so–called, is part of the middle distillate section of the crude distillation column. The generic name for it is gas oil and for road vehicles it is known as automotive gas oil (AGO). AGO is a more complex and variable fuel than petroleum. Its density can vary considerably, as can its volatility and general composition. Europe rather than the USA is likely to set the future trends for AGO quality due to the size of the market and the much greater importance of the diesel passenger car.

2.3.2 Important diesel fuel properties

The importance of diesel fuel properties such as density, volatility, sulphur, aromatics and olefin content have been carefully studied. Primary fuel factors that affect particulate emissions are volatility, sulphur and aromatic content, with sulphur the dominant factor according to Egebäck and Westerholm (1992), quoting Wall and Hoekman (1984). Extensive testing in Sweden involving the Swedish Environmental Protection Agency and other organisations, using a truck and bus with reformulated diesel fuel have shown that decreasing the volatility (increasing the T_{10} temperature) and reducing sulphur both have a significant emissions benefit, but that the effect of reducing aromatics was more beneficial to emissions.

Increasing the fuel's cetane number (CN) can also provide a similar emissions reduction benefit to that of reducing sulphur or volatility. Increasing the CN shortens ignition delay, and effectively increases flammability. Brandt (1991) confirms that increased CN can reduce HC, CO and particulate emissions, with no rise in NO_x, as demonstrated during transient testing of a 1991–standard US diesel engine.

The European Motor Industry Association, ACEA, has put forward a proposal for a future AGO quality specification to assist in meeting proposed and anticipated (more) stringent engine emission standards. The ACEA proposes a narrow density range (810–840 km/m³), a high cetane number (58), low sulphur content (0.02% mass), low aromatic content (10% volume) and a reduced distillation range (fewer heavier fractions). While the petrochemical industry acknowledges that such a fuel specification is technically possible, in practice it is considered infeasible in the volumes required for Europe (Rivers *et al*, 1993). Implications for some of these diesel fuel properties are discussed in the following sections.

2.3.2.1 Sulphur

The combustion of diesel fuel produces larger quantities of oxides of sulphur (SO_x) than does petrol. SO_x formation is a function of the sulphur content of the fuel, which can be 10 times higher for diesel than for petrol. Legislation in the US and Europe is addressing this issue and European diesel, from October 1994, will have a maximum sulphur content of 0.2 percent by weight. From October 1996, the limit will be 0.05 percent sulphur by weight.

Refineries use hydrodesulphurisation plant to remove sulphur from fuel, but it is anticipated that 0.05% sulphur diesel can only be partly achieved by modification of existing plant. Rivers *et al* (1993) predict that the reduction to 0.05% could require between 28 and 44 Mt per year of additional hydrodesulphurisation capacity in the EC by 1996. In many refineries investment in new equipment will be required (estimated for the EC at between US$3,300 and 5,000M) and the energy to remove sulphur to low levels is significant, leading to a possible 5 percent (4.6 Mt per year) increase in CO_2 emissions from EC refineries (Bruner, 1991). Commercial Motor (1989) estimates that diesel fuel prices will rise by about three percent to achieve the 0.05% sulphur limit.

2.3.2.2   Cetane number

The cetane quality of AGO can be improved by removing poor quality conversion streams, but this would immediately reduce AGO quantities since only the top tier of the straight run AGO components have a sufficiently high cetane number. Figure 8 (from Rivers *et al*, 1993) shows the quality (cetane number and density) of AGO blending components.

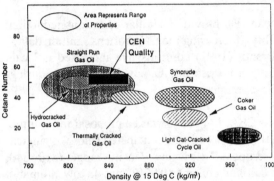

The cetane can also be raised by switching to more paraffinic crudes, but produce poorer petrol components. The use of additives (alkyl nitrates) are effective ways of improving cetane quality and are seen by the refineries as a more attractive way of meeting any future specification enhancement for cetane (Rivers *et al*, 1993).

Figure 8. AGO blending components

Another way of improving the cetane number (and reducing density) of AGO with little yield loss is by hydrogenation. This involves the conversion of aromatics to napthenes either at very high pressures over a single catalyst or in two stages involving different catalysts. The most dramatic improvements in quality can be obtained with the poorest quality feedstocks where cetane numbers can be improved by up to 25 points (Rivers *et al*, 1993). The financial costs are high, as is the energy required to manufacture the hydrogen to saturate the aromatics. Capital investment for hydrogenating light cycle oils in the EC has been estimated at between US$2,400 and 6,000M and the refinery CO_2 emissions are predicted to rise by over 8 Mt per year (ibid).

2.3.2.3   Density

Lower AGO density can lead to reduced particulate emissions in some light duty vehicles, but the response of heavy duty engines to density changes is heavily dependent on engine design. Heavy duty engine particulate emissions are reported as decreasing, remaining constant or even increasing with decreasing fuel density across a range of engines (Rivers *et al*, 1993). Lower density AGO could be produced by removing the higher boiling point material, but would reduce AGO availability (refer to Figure 8) and lower cetane quality thereby leading to a loss of engine power and fuel economy. Alternatively, kerosine can be added to gasoil to reduce density, but would increase the AGO supply at the expense of Jet fuel which is in growing demand.

2.4   Commercially available reformulated fuels

In addition to the reformulated gasolines available in the US (such as Phillips Petroleum *Unleaded Plus* – see section **2.2.3.2**), several "clean" petrol and diesel fuels are now being marketed in the EC.

Greenergy (Sweden) AB is the largest independent manufacturer and supplier of environmental fuels in Scandinavia. Greenergy DMF diesel fuel has been available in Sweden since 1989 and is claimed to account for over three–quarters of the total Swedish diesel market. Its reformulation includes 0.001% sulphur, CN of 55–58, density of 820 kg/m^3, 12% aromatics and a reduced distillation range (190–320°C). Some of the claimed emission benefits from DMF diesel include a 98 percent reduction in SO_x, 50 percent reduction in particulates and a 20 percent NO_x reduction. Greenergy estimate that existing refinery infrastructure could supply 10 percent (10 Mt per annum) of the current EC diesel demand with reformulated fuel like DMF. The company claims the cost of the product (excluding taxes) to be 4 pence/litre more than conventional diesel fuel, and have recently (during 1993) commenced UK operations with the intention to be able to supply to anywhere in the UK by the end of 1994 (Greenergy, *undated*).

The Finnish oil company, Neste Oy, has developed a range of advanced, reformulated petroleum products such as City Diesel and City Gasoline. Demand for lower–emission fuels is increasing in a number of European countries, particularly in the Nordic region, and Neste was the first European oil company to launch a reformulated petrol, City Gasoline, in May 1991. It is claimed that vehicles using the reformulated petrol release 15 to 20 percent less CO and about 15 percent fewer reactive HCs than with conventional fuel. Vehicles fuelled with City Diesel are said to emit 95 percent fewer sulphur compounds, nearly 20 percent less CO and up to 15 percent less NO_x. The City Diesel also allows modern vehicles to use particulate traps and oxidising catalysts more effectively (ibid).

2.5   Synthetic liquid fuels

The term "synthetic fuel" is generally taken to mean any fuel which is made from a feedstock by a process of chemical conversion. The main feedstocks available for conversion to synthetic fuel are:

- residual fuel oil from crude oil refining;

- unconventional oil sources such as very heavy crudes, tar sands and oil shales;

- coal;

- natural gas and petroleum gas; and

- biomass in the form of organic wastes or crops grown specifically for energy purposes.

In addition, electricity from nuclear power or renewable sources may be considered as a synthetic resource which can be utilised by electrochemical means.

Langley (1987) reports for the Department of Energy on the relative technical and economic merits of various synthetic fuels for road transport applications in the UK. Due to compatibility requirements with existing engines, at least in the short to medium term and the constraints imposed by the infrastructure for fuel distribution, any synthetic fuel considered viable for the future must be able to be directly substituted for conventional petrol and diesel fuels.

All of the feedstocks listed (except for nuclear/renewable electricity) can be converted to liquid hydrocarbon fuels which have similar characteristics to conventional petrol and diesel. The extent of the processing required and the yield depends mostly on the nature of the feedstock. Some of the fuels, such as methanol converted from natural gas, ethanol distilled from biomass and electricity, are discussed in the relevant chapters later in this book. This section discusses the less conventional synthetic liquid fuels, such as those from coal, tar sands and oil shales (a sedimentary rock formed by the deposition of silt and organic matter in shallow freshwater lakes. The organic matter becomes transformed by geological processes to a polymeric hydrocarbon material known as *kerogen* which, when heated to 480°C, decomposes to release shale oil whose composition is similar to heavy crude oil but with a much higher nitrogen content).

The main options for synthetic fuels derived from these feedstocks can be summarised as follows:

- **unconventional oil sources,** which include very heavy crude oil, tar sands and oil shale, can be converted to products matching current petrol and diesel fuel specifications by means of existing thermal cracking (a process which breaks down heavy hydrocarbon fractions into lighter ones by heating the feedstock) or hydrocracking (which involves the addition of hydrogen in the presence of a catalyst);

- **coal** can be converted to liquid fuels indirectly by gasification (in which hydrogen is added as steam while carbon is removed as CO_2) followed by a number of synthetic routes to produce either methanol or a synthetic hydrocarbon fuel (petrol or diesel). Alternatively, direct liquefaction methods (in which hydrogen is added either in gaseous form or via a hydrogen–rich donor solvent) can be used to produce synthetic hydrocarbon fuels.

The choice of process for converting feedstocks to synthetic fuels depends largely on the nature of the available feedstock and the required product. The feedstocks considered here have two characteristics in common to varying degree – they are deficient in hydrogen and contain undesirable material such as ash and sulphur. Coal is particularly deficient in hydrogen (with 0.6 to 0.8 hydrogen atoms for each carbon atom – petrol has 1.8) and oil shales and tar sands can contain over 80 percent ash.

The Department of Energy study (Langley, 1987) concluded that when linking resource constraints and infrastructure costs with the product costs, a ranking of the three most viable options for synthetic fuel production from indigenous resources within the UK is as follows:

i)   petrol from fuel oil or very heavy crude;

ii) petrol from coal (by direct liquefaction);

iii) kerosine/gasoil from coal (Shell gasifier/Middle Distillate Synthesis).

Only the first of the three options listed was considered to be economically viable in the short term, but the remaining options could progressively become viable in the future. It was also considered that imported fuels from lower cost overseas sources were likely to be more economic than domestically produced synthetic fuels in most future scenarios.

New Scientist (1993) reports that enzymes, derived from bacteria that live in the human gut, that are able to break down coal into oil. The Oak Ridge National Laboratory in Tennessee has developed a system that can liquefy as much as two–fifths of the coal fed into it, with the solid residue additionally able to be used as a fuel. The advantage over conventional coal liquefaction systems (that require temperatures of up to 500°C and pressures of several thousand atmospheres) is low temperature (less than 40°C) and atmospheric pressure.

The coal is granulated (to 100 microns) and placed in a tapered column. The enzymes (called hydrogenases) are dissolved in an organic solvent such as benzene, and squirted upwards through the coal particles together with hydrogen. The enzymes catalyse a reduction reaction in which the coal reacts with the hydrogen to form a liquid – this oil dissolves in the organic solvent and can then be extracted. So far, bituminous coal has worked best, but reaction vessels no larger than half–a–litre have yet been used for the process. Commercialisation of this system, therefore, is viewed with caution, and is not expected before 2005 (ibid).

2.6   The future for liquid hydrocarbon fuels

As matters stand today on the world oil market, there is no reason to believe that crude oil with its many and varied products and byproducts will not be one of the most competitively–priced sources of energy for a considerable period of time. Petrol and diesel fuels will be further developed and used in engines well into the next century.

Fuel composition has a role to play in reducing exhaust emissions, although a life cycle analysis should be undertaken in determining how fuels are changed. For example, changing a fuel specification to achieve lower vehicle exhaust emissions may be offset by an increase in processing energy (and emissions produced elsewhere) – such as with diesel fuel desulphurisation. Certain vehicle technology classes are more likely to benefit from reformulated fuels than others – uncontrolled petrol engines (which will form the majority of the EC passenger car fleet until about 1996/97 and will still represent 25% in 2002) are likely to demonstrate higher emissions reduction than cars with three–way catalysts.

It is probable that Europe will adopt certain fuel reformulations sometime in the future, although not necessarily in line with the present US proposals. Certain changes in fuel specification have already been mandated, such as diesel fuel sulphur content. Due to potential health concerns, the limitation of aromatic content (for petrol and diesel) and the reduction of benzene in petrol is likely, especially since average EC levels are currently much higher (by about 50 percent) than in the US (ARCO, 1993). Good octane is likely to be maintained by the addition of oxygenates, with a consequent beneficial reduction in exhaust CO and in the quantity and reactivity of HC emissions.

The use of oxygenates as high octane blending components is increasing, as demonstrated in the USA. Methanol and ethanol, in low volume, are also frequently blended with petrol, usually as an extender, but since they contain high levels of oxygen they reduce the calorific value of the fuel and thereby worsen fuel economy. The alcohols also have high vapour pressures which, when blended with petrol, increase the overall volatility of the fuel and hence evaporative emissions. Two of the key oxygenates that are promoted for use in petrol are MTBE and ETBE since they have sufficiently low vapour pressures, high octane and a reasonable oxygen content. Ethanol, the main oxygenate competitor to MTBE, is distilled from grain (and can be produced from other biomass) and has advantages as a "renewable" additive, helping to extend conventional fuel supplies.

A reduction in evaporative emissions would be achieved through lowering the fuels' vapour pressure, which is generally 10 percent higher in the EC than the US (ibid). The sulphur content of petrol has a major effect on NO_x emissions (AECC, 1994) with any reduction enabling the three–way catalytic converter to function more efficiently. However, current EC petrol contains about 30 percent less sulphur than US fuel (ARCO, 1993), so any reduction to meet future EC specifications may have fewer implications than for US refineries, depending upon the level of sulphur stipulated.

In addition to lowering the sulphur content of diesel fuel, increasing the cetane rating has been demonstrated as an effective way of reducing exhaust emissions. Other, less well proven means of reducing diesel emissions include reducing the fuel density and distillation range, but given the oil industry's concerns about the implications for reduced supply, this route seems less likely.

In two or three decades crude oil prices are likely to have risen significantly in real terms, especially if global petrochemical demand continues to increase and developing countries aspire to the standard of living and level of personal mobility enjoyed in the West. Other forms of liquid hydrocarbon fuels, derived from unconventional sources, are then more likely to become economical to produce. The most viable form of synthetic fuel able to be introduced is petrol from fuel oil or very heavy crude and, if crude oil prices rise substantially, petrol from coal and kerosine/gasoil from coal.

2.7 Summary

Reformulated petrol and oxygenates (results from US AQIRP)

+ MTBE $\Rightarrow$ ↓ CO, ↓ HC reactivity, ↑ formaldehyde, can ↑ NO_x
↓ aromatics (lower benzene) $\Rightarrow$ ↓ CO, ↓ HC, ↓ benzene
↓ olefins $\Rightarrow$ ↓ NO_x, ↓ 1,3–butadiene, ↑ HC (but ↓ reactivity)
↓ sulphur $\Rightarrow$ ↓ CO, ↓ HC, ↓ NO_x, ↓ toxics
↓ T_{90} $\Rightarrow$ ↓ HC, ↓ toxics
↓ RVP $\Rightarrow$ ↓ evap. emissions (diurnal more than hot soak)

Oxygenates $\Rightarrow$ ↓ CO, ↓ HC reactivity, may ↑ NO_x, may ↑ aldehydes

Alcohols can reduce certain emissions, and lead to lower HC reactivities, but the lower calorific values worsen fuel economy and the higher vapour pressures can increase evap. emissions. Ethers, such as ETBE and especially MTBE, with high octane and lower RVP (than alcohols) appear to be the most favoured oxygenates for the future. Reformulated fuels are likely to be best employed in niche markets, targeting urban pollution concerns and may provide the most significant benefits in uncontrolled petrol cars.

Reformulated/low sulphur diesel

↑ cetane number, ↓ sulphur (and, to a lesser extent, ↓ distillation range and ↓ density) $\Rightarrow$ significant reductions in regulated emissions (mostly CO, HC and particulates).

Reformulated and low sulphur diesel (as with reformulated petrol) will mean the conversion of certain less desirable fuel components into other blending components. This involves energy and causes emissions (as does the removal of sulphur) which should be assessed against the local air quality improvements that may arise from the fuels' use.

REFERENCES FOR CHAPTER 2

ACEA/EUROPIA (1993). Press Release: European Auto/Oil Research Programme launched. ACEA/EUROPIA, Brussels, 19 July 1993.

AECC NEWSLETTER (1994). North America: Details of Reformulated Gasoline Regulation. Pages 6–7, January 1994. Automobile Emissions Control by Catalyst, Brussels.

ARCO CHEMICAL EUROPE (1993). European Clean Gasoline. ARCO Chemical Europe, Maidenhead, Berkshire.

BENSON J D, V R BURNS, R A GORSE Jr, A M HOCHHAUSER, W J KÖHL, L J PAINTER and R M REUTER (1991). Effects of Gasoline Sulfur Level on Mass Exhaust Emissions – Auto/Oil Air Quality Improvement Research Program. SAE Technical Paper Series No 912323, Society of Automotive Engineers, Inc., Warrendale, Pennsylvania, United States of America.

BRANDT G (1991). Fuel additives – towards a cleaner environment. Proceedings from the conference "Engine and Environment: Which fuel for the future?", 23–24 July 1991, Grazer Congress, Graz, Austria.

BRUNER I (1991). The complex environment of oil refineries. Proceedings from the conference "Engine and Environment: Which fuel for the future?", 23–24 July 1991, Grazer Congress, Graz, Austria.

COLUCCI J M and J J WISE (1992). Auto/Oil Air Quality Improvement Research Program – what is it and what has it learned? Proceedings from the XXIV FISITA Congress "Automotive Technology Serving Society", 7–11 June 1992, Institution of Mechanical Engineers, London.

COMMERCIAL MOTOR (1989). Engineer's notebook: Exhaustive research. Pages 44–45, 28 September–4 October 1989.

COORDINATING RESEARCH COUNCIL (1993). Auto/Oil Air Quality Improvement Research Program – Phase I Final Report. May 1993. CRC Inc., Atlanta, Georgia, United States of America.

CRAGG C (1992). Cleaning up motor car pollution. New fuels and technology. Financial Times Management Report, Financial Times Business Information, London.

EGEBÄCK K K and R WESTERHOLM (1992). Impact of diesel fuels on exhaust emissions. Proceedings from the XXIV FISITA Congress "Automotive Technology Serving Society", 7–11 June 1992, Institution of Mechanical Engineers, London.

GORSE Jr R A, J D BENSON, V R BURNS, A M HOCHHAUSER, W J KÖHL, L J PAINTER, R M REUTER and B H RIPPON (1991). Toxic Air Pollutant Exhaust Emissions with Reformulated Gasolines. SAE Technical Paper Series No 912324, Society of Automotive Engineers, Inc., Warrendale, Pennsylvania, United States of America.

GREENERGY (Undated). Greenergy (Sweden) AB – an environmental fuels company. Press Release.

HOCHHAUSER A M, J D BENSON, V R BURNS, R A GORSE Jr, W J KÖHL, L J PAINTER, R M REUTER, B H RIPPON and J A RUTHERFORD (1991). The Effect of Aromatics, MTBE, Olefins, and T_{90} on Mass Exhaust Emissions from Current and Older Vehicles – The Auto/Oil Air Quality Improvement Research Program. SAE Technical Paper Series No 912322, Society of Automotive Engineers, Inc., Warrendale, Pennsylvania, United States of America.

LANGLEY K F (1987). A ranking of synthetic fuel options for road transport applications in the UK. ETSU–R–33. Energy Technology Support Unit, Harwell.

NEW SCIENTIST (1993). Technology: Gut power fuels cheaper coal oil. Page 20, 30 January 1993.

OFFICIAL JOURNAL OF THE EUROPEAN COMMUNITIES (1985). Council Directive of 5 December 1985 on crude–oil savings through the use of substitute fuel components in petrol (85/536/EEC). OJ No L334/20, 12.12.1985.

PETROLEUM REVIEW (1993). Expansion in US regulation of gasoline emissions. Pages 229–231, Petroleum Review, May 1993, The Institute of Petroleum.

REUTER R M, J D BENSON, V R BURNS, R A GORSE Jr, A M HOCHHAUSER, W J KÖHL, L J PAINTER, B H RIPPON and J A RUTHERFORD (1992). Effects of Oxygenated Fuels and RVP on Automotive Emissions – Auto/Oil Air Quality Improvement Program. SAE Technical Paper Series No 920326, Society of Automotive Engineers, Inc., Warrendale, Pennsylvania, United States of America.

RIVERS K J, C W C PAASSEN, M BOOTH and J M MARRIOTT (1993). Future diesel fuel quality – balancing requirements. Proceedings from the seminar "Worldwide engine emission standards and how to meet them", 25–26 May 1993, Institution of Mechanical Engineers, London.

SCHOONVELD G A and W F MARSHALL (1991). The Total Effect of a Reformulated Gasoline on Vehicle Emissions by Technology (1973 to 1989). SAE Technical Paper Series No 910380, Society of Automotive Engineers, Inc., Warrendale, Pennsylvania, United States of America.

US CONGRESS (1990). Replacing gasoline: Alternative fuels for light–duty vehicles, OTA–E–364. US Congress, Office of Technology Assessment, Washington, DC: US Government Printing Office.

WALL J C and S K HOEKMAN (1984). Fuel composition effects on heavy–duty diesel particulate emissions. SAE Technical Paper Series No 841364, Society of Automotive Engineers, Inc., Warrendale, Pennsylvania, United States of America.

WILLIAMS D and M W VINCENT (1992). Past and anticipated changes in European gasoline octane quality and vehicle performance. Proceedings from the XXIV FISITA Congress "Automotive Technology Serving Society", 7–11 June 1992, Institution of Mechanical Engineers, London.

3.   METHANOL

3.1   Introduction

Interest in using alcohols, and especially methanol, as an alternative transport fuel first began in the United States, initially because it was seen as a relatively cheap, non-petroleum based fuel which offered the possibility of reduced dependence on imported oil (Holman *et al*, 1991). More recently it has been seen as a shorter term replacement for petrol to improve urban air quality and has been given prominence by the proposed amendments to the Clean Air Act (Waters, 1992). One aspect is the call for 500,000 "clean fuel technology" cars to have been sold in the US by 1995, and, although the Bill defines "clean fuel" as methanol, ethanol, propanol or electricity, the generally accepted view is that methanol is likely to be the most practical way of meeting this target.

Methanol can be used as a fuel by itself (known as M100) or mixed with petrol in differing quantities, the most common being 85 per cent methanol and 15 per cent petrol, commonly referred to as M85. Flexible fuelled vehicles are being developed which can run on either methanol or petrol as a short-term transitional solution to the current sparsely distributed methanol refuelling stations.

One advantage methanol has over some other alternative fuels is that it is a liquid at normal room temperature, allowing it to be stored in conventional fuel tanks. It can be used in a conventional passenger car petrol engine and also in diesel engines with spark (or ignition) assistance. Its main disadvantages are that it is corrosive and highly toxic and has a lower energy density than petrol, leading to a smaller vehicle range with a current fuel tank.

3.2   Fuel characteristics

The properties of alcohol fuels are favourable to improved efficiency of the Otto (spark-ignition petrol engine) cycle. In this cycle, efficiency increases with increased compression ratio, with increased leaning of the air/fuel mixture, and with high flame speed. Methanol possesses a high octane rating - its Research Octane Number, or RON is 111 (Zelenka *et al*, 1991), has broad flammability limits for smooth lean operation and high flame speed. In addition, the high latent heat of vaporisation provides for an increase in volumetric efficiency by decreasing the work required for compression of the mixture and increasing the charge density.

These properties lead to a higher engine thermal efficiency - less fuel energy is therefore used per unit of work output. The theoretical thermal efficiency is 10-25 percent higher than with petrol, although demonstrated increases have normally been under 10 percent. A higher power output is additionally obtained from a methanol-fuelled engine of similar displacement to a petrol version; smaller capacity units may therefore be used for equivalent power. With early prototype methanol-fuelled cars (using M85-M90) power increases have been demonstrated at between four and 22 percent (DeLuchi *et al*, 1988a). However, increased efficiency gains are offset, to some degree, by increased vehicle weight because of the larger (and heavier) fuel tank.

Methanol has a lower energy density than petrol due to the high atomic weight of the oxygen it contains. On a mass basis petrol provides over twice the heating energy of methanol, but because of the advantage of increased engine thermal efficiency, the volume of methanol is currently around 1.8 times that of petrol required for equivalent vehicle range. Dedicated methanol engines will exhibit even higher thermal efficiencies than modified petrol engines or flexible fuelled vehicles, and it is expected that in the foreseeable future methanol vehicles will use 1½ times the volume of petrol in conventional vehicles (Cragg, 1992).

Methanol has a low cetane rating, making it difficult to use in a diesel (compression-ignition) engine without spark or ignition assistance or a fuel additive. Its low vapour pressure, while reducing evaporative losses, makes cold starting more difficult unless fuel additives (such as petrol), direct fuel injection or mixture heating are employed. Cold start combustion in neat methanol-fuelled engines can also be improved by means of exhaust gas recirculation (Gardiner *et al*, 1991). Another method to aid cold starting and minimise "cold" engine exhaust emissions is the addition of propane or butane to the fuel (Lee *et al*, 1987).

Methanol is corrosive, and certain alloys and polymers commonly used in the engine and fuel systems of vehicles need to be replaced. Certain lubricants may need to be changed or specially formulated, and lubrication of certain components such as the fuel pump or injection pump would be necessary where methanol replaces diesel as a fuel. Alcohols are hygroscopic (attract and absorb water), and while this would enable methanol fires to be extinguished with water, the fuel requires careful dry handling and storage. Corrosion inhibitors may need to be added to the fuel.

3.2.1 Safety implications

Methanol is colourless and has no taste or odour and is highly toxic. It can be absorbed through the skin more quickly than petrol (US Congress, 1990) and cause blindness and death by ingestion of between 25 and 100 millilitres, generally a smaller quantity than death by ingestion of petrol. However, the risks of ingestion or skin exposure can be minimised with appropriate design of refuelling systems and the fact that the use of methanol in household or garden applications (such as lawnmowers) is unlikely. No operational problems, safety or otherwise, were encountered with any refuelling during a six-year US Army methanol fleet trial, covering over a million vehicle miles using 64 light-duty vehicles (Baber *et al*, 1990).

On the subject of methanol fires, the flame has low luminosity, making it difficult, if not impossible to see in daylight without the addition of a marker–tracer. While methanol is unlikely to ignite in the open air after a spill (lower flammability limit is 6.7%, compared with petrol's 1.4% by volume, in air), neat fuel vapour (but not M85) can explode in a closed fuel tank if enough energy is supplied, due to a much higher upper flammability limit of 36.0% compared with 7.6% for petrol. Bladder–type fuel tanks, which avoid creating an air space as the tank empties, may be necessary for M100 vehicles, or alternatively flame arrestors could be located at the fuel tank mouth.

3.3   Feedstocks

Methanol can be manufactured from coal, natural gas or biomass. In the UK, coal is abundant and would appear to offer energy security in the long term. However, in commercial methanol production, coal is not the preferred feedstock because the conversion process is complex, costly and more energy–intensive than using other feedstocks. Natural gas is the cheapest feedstock for methanol conversion (Holman *et al*, 1991), and although the conversion process is more energy efficient than that from coal, there is a cost in energy terms relative to using the natural gas directly as a vehicle fuel itself. Currently the most effective process of producing methanol from natural gas is the Lurgi combined reforming and low–pressure synthesis, using air and water in addition to the natural gas itself (Volkswagen AG, undated).

The idea of using methanol produced from natural gas as a transport fuel, first seriously discussed in the 1970s, was because it was seen as a means to obtain the emissions advantages of natural gas by converting methane into a more practical liquid fuel.

Natural gas reserves are more widely distributed than oil, as demonstrated in Tables 1 and 2 in chapter 1. While the Middle East possesses two–thirds of the known global oil reserves, it only has one–third of the gas. Western Europe and the former Soviet Union can only claim to have 7½ percent of the global oil reserves, but almost 45 percent of the known natural gas. This prospect places Europe in a favourable position for using natural gas as a feedstock for methanol, with considerable reserves close to hand and the prospect of more reliable distribution and consequent increased energy security and fuel diversification.

A proportion of the methane used for methanol conversion can be obtained from flared gas. Several studies have concluded that substantial quantities of methanol, between 1.4 and 4 million barrels per day (64–182 million tonnes per year), can be produced from this source (IEA, 1990). However, certain remote locations may preclude the use of flared gas as a methanol feedstock due to economic considerations. Additionally, the amount of flared gas has fallen from 173 billion cubic metres in 1975 to 92 bcm in 1988, with the excess gas reinjected for other uses (Cedigaz, 1989). Nevertheless, a considerable supply of low–cost flared natural gas is potentially available for methanol conversion which poses

fewer demands on the market for conventional natural gas reserves, especially as many
new applications are now being developed for the fuel (such as combined cycle gas
turbine power stations) which will tend to force up the price in future years.

3.4   Infrastructure

Worldwide methanol demand in 1989 was 18.2 million tonnes, primarily for chemical uses
(Cragg, 1992). Declared capacity of the methanol production industry revealed that
surplus capacity was 2 million tonnes, which, when expressed in petrol equivalent, is
about 1.1 million tonnes – or 0.15 percent of the 1989 world petrol demand. Clearly, any
significant increase in the use of methanol as a vehicle fuel requires greatly expanded
methanol production capacity. For example, if existing capacity were to triple and the
demand for methanol as a chemical feedstock remained static, there would be enough
methanol to fuel just 3 percent of the world's petrol–fuelled vehicles. While there has
been a reasonable increase in methanol plant capacity in recent years, much of the
increase is based on assumptions about the growth of the chemical market for methanol
and, in particular, as a feedstock for the expected expansion in the MTBE (methyl tertiary
butyl ether) requirement (an octane–raising oxygenate) for reformulated gasoline.

Modern oil refineries operate at around 93% efficiency, having had over 100 years to
develop the refining process. The efficiency of plant converting methane into methanol
is in the range 60–75 percent in energy terms. Oil refineries are also highly flexible in
what they produce – the product mix can be altered to suit the market. A methanol plant
produces a single product with corresponding greater financial risks as demonstrated by
the varying fortunes of existing methanol manufacturers.

The distribution, storage and refuelling infrastructure for methanol is more costly and
complex than that for petrol. A greater volume of methanol (1.8, falling to perhaps 1.5
times) would need to be distributed for the equivalent petrol it replaces, meaning that
extra fuel distribution tanker capacity and storage tank volume are necessary. The storage
containers and tanks need to be constructed of, or lined with, methanol–resistant materials,
and systems to ensure dry storage would need to be employed to prevent water absorption.
Filling stations would need new tanks, pipes and pumps, and the method of refuelling
would need to ensure no spill, splashback or leakage of the fuel occurred.

3.5   Vehicle modifications

3.5.1 Petrol–engined vehicles

Methanol can be used as a vehicle fuel in the petrol engine with very little modification,
except to replace the components susceptible to corrosion by contact with the fuel.
Volumetric efficiency will be increased due to methanol's high heat of vaporisation, but
to achieve the highest increase in thermal efficiency the compression ratio should be

raised in line with methanol's high octane rating. This can lead to an efficiency increase of up to 15 percent over a petrol engine. A vehicle running today on methanol with a fuel tank similar in volume to a petrol–engined vehicle will return a range of about 55 percent of that of the petrol vehicle. With a dedicated methanol engine benefitting from higher efficiency gains, the range is likely to rise to 67 percent. Put another way, a methanol fuel tank is expected to be 1½ times the volume of a petrol tank for equivalent vehicle range with a dedicated engine.

Lubricants must be formulated specifically for the use of methanol in Otto engines. Oils adapted to alcohol fuels by modified additive packages have been developed and used in both commercial and demonstration alcohol car fleets in Brazil and California. Anti-corrosion additives may also help to reduce engine wear. Development work to combat corrosion and wear is still ongoing, but it appears that engine durability over 100,000 miles of operation is achievable with properly specified fuels and oil formulations. The addition of petrol at 15 percent (M85 blend) has demonstrated an ability to improve lubricity and reduce wear, in addition to improving cold starting and cold driveability.

Figure 9 (from Volkswagen AG, undated) shows the VW multi–fuel concept (MFC) for variable methanol/petrol operation up to M90, developed specifically for California. The figure shows which components require modification and identifies new items, such as special lubricant formulation and the requirement for an alcohol sensor.

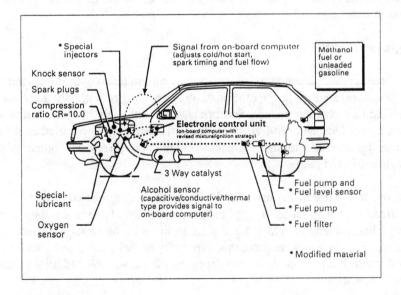

Figure 9.   Schematic diagram of VW multi–fuel concept

Flexible fuel vehicles that can run on methanol/petrol blends, usually in any proportion (although most FFVs developed in the United States have been capable of running from all–petrol to M85, but not neat methanol due to cold starting problems), are seen as a way of introducing alternative fuels into the market, even without a full refuelling infrastructure in place. This technology was originally developed by Ford in conjunction with TNO of the Netherlands and has been made more reliable by the development of electronic fuel injection. An optical fuel sensor determines the alcohol content and the electronic control system automatically adjusts the fuel injection system and ignition timing. However, the system cannot be optimised since the compression ratio is set for the petrol octane rating, and not the higher value of the methanol.

Another means of using methanol as a vehicle fuel is in the dissociated methanol engine. A certain amount of research and development has gone into this concept (Sakai *et al*, 1987 and Seppen & Ter Rele, 1989), but vehicle systems which use dissociated methanol engines have not been extensively promoted because of the complexity in controlling them. The principle is based on energy recovery from exhaust heat which promotes a catalytic reaction in the exhaust dissociator that converts methanol into carbon monoxide and hydrogen ($CH_3OH \rightarrow CO + 2H_2$). This dissociated gas is added to the methanol and the combined fuel, with a heating value increased by more than 20 percent (due to the hydrogen), leads to greater thermal efficiency and a fuel saving of around 12 percent in practice (in prototype engines). The addition of hydrogen allows lean combustion, lowering exhaust emissions relative to neat methanol, although no published test results have been available to date.

3.5.2 Diesel–engined vehicles

The function of a diesel engine depends to a large degree on the auto–ignition property (cetane rating) of the fuel employed. Unfortunately methanol, as with all alcohols, displays poor auto–ignition characteristics. But because alcohol fuels burn with lower NO_x formation than diesel fuel, and with virtually no particulate emissions, methanol is seen as an attractive option for fuelling the diesel engine, especially as comparable efficiency is obtainable (IEA, 1990).

The technical solutions to the use of methanol in the diesel engine are of two basic types. One involves modification of the fuel itself, with only minor engine changes and the other requires more comprehensive engine and vehicle adaptation and modification. Also, as mentioned previously, lubricity and wear problems, especially in fuel pumps and injectors, and corrosion need careful attention with the development of suitable additives and specially formulated lubricants.

Methanol can be modified for use in the diesel engine by two ways. One is by the addition of ignition–improving agents and the other is through emulsification of the alcohol fuel in diesel oil, but does not enable complete substitution for diesel fuel.

Ignition–improving agents, such as ICI's "Avocet", provide a means of allowing diesel engines to operate on methanol when the combustion air is not well preheated. Once a diesel is started, it can operate on pure alcohols under high–speed, high–load conditions, but cold–start, part–load and transient operation is not possible due to misfiring as the engine temperature falls. Quite high levels of around 5 percent (or higher in cold climates) of costly ignition improvers have to be used for reliable diesel engine operation.

The ways in which the diesel engine can be modified and adapted to allow the use of straight alcohols in spite of their poor self–ignition properties include:

- A two–fuel system of partial replacement of the diesel fuel by alcohol carburation (or "fumigation") of the air inducted to the engine;

- Duplication of the injection system, one for diesel fuel as a pilot injection to initiate combustion and one for the methanol as main fuel (dual–fuel); and

- The ignition of injected methanol by spark plugs or glow plugs for the complete replacement of the diesel fuel.

It can be argued that the spark–assisted engine is no longer a diesel, since the ignition is not by compression, but the cycle is similar to the diesel in that fuel injection continues after the initiation of the combustion by the spark. As in the diesel engine, the energy release will, therefore, be controlled by the injection timing.

Materials problems have been encountered with fumigation in turbocharged engines due to rapid erosion of the compressor blades caused by liquid droplets of methanol. Excessive piston and piston ring wear have also been observed, making fumigation an unpromising technique at this stage.

The injection of two fuels into the diesel engine combustion chamber has proven very suitable for methanol usage if fuel pumps and injectors are externally lubricated or redesigned to compensate for the poor lubricity of alcohols. The system has the potential for very high replacement of the diesel fuel, from 75–90 percent depending on the duty cycle. In fact a 100 percent replacement would be possible if the high–cetane pilot fuel consisted of methanol with a sufficient content of ignition improver to burn a number of fuels with low ignitability as secondary or main energy fuel (e.g. water–containing alcohols). The drawback of the dual–fuel, dual–injection concept is the increased cost of a second separate fuel system. The thermal efficiency has been shown to be as good as, or better than, the diesel counterpart.

Alcohols, particularly methanol, may ignite on hot surfaces. This pre–ignition property is a limitation in Otto engines, but it can be turned into an advantage for direct–fuel–injection diesel engines by the use of glow plugs for the normal ignition when contacted by fuel vapour. Field testing of both two and four–stroke diesel engines adapted to methanol with glow plug ignition assistance has been underway for some years. Zelenka *et al* (1991) report the successful development of a 2.3–litre direct–injection diesel engine modified for methanol fuelling by glow plug assistance.

There is also interest in research into compression–ignition engines which operate on methanol at higher temperatures, utilising the heat–retention properties of ceramic engine components and Zelenka *et al* also report successful compression ignition with methanol achieved with a higher compression ratio than usual (24:1 instead of 20:1) in a single–cylinder research engine. Johns *et al* (1991) report that the Detroit Diesel Corporation's 6V–92TA two–stroke diesel engine operates on methanol in the mid to full load range without ignition aids, by employing a higher compression ratio and reducing the scavenging to retain a higher level of hot residual exhaust gases in the cylinders.

3.6   Emissions performance

3.6.1   Vehicle exhaust emissions

3.6.1.1   Substituted for petrol

Methanol has a lower carbon to hydrogen ratio than petrol (and diesel fuel), and therefore produces less carbon dioxide when burnt. On an equivalent energy basis, methanol produces 93% of the CO_2 emitted from the combustion of petrol (see Table 25 in chapter 10). Because the use of methanol improves engine efficiency, on a per unit distance travelled basis, allowing for a 15 percent efficiency improvement, methanol produces around 81% of the CO_2 of a petrol vehicle. However, account should be taken of the extra volume and weight of fuel, and its storage system (for equivalent vehicle range), which will offset some of the overall efficiency advantage.

Because alcohols have a greater internal cooling effect, owing to their high latent heat of vaporisation, and because the fuel burns with a lower flame temperature than for petrol, less NO_x formation occurs. However, if the compression ratio is increased to take advantage of methanol's higher octane rating, NO_x emissions will increase. If methanol is used in a modified engine or in a flexible fuel vehicle where compression is unchanged, a higher NO_x emissions advantage will be realised than for a dedicated methanol engine (IEA, 1990).

The emissions of regulated pollutants from methanol–fuelled vehicles appear to be variable and some test results from fleet trials have proved ambiguous. The main advantage would appear to arise from lower evaporative emissions – especially with neat methanol as a fuel (Cragg, 1992). Certain experience from FFVs using M85 in the United States suggests that average non–methane HC exhaust emissions are about one–third higher than from 1989–certified, three–way catalyst–equipped petrol–fuelled cars (US

Congress, 1992). Although emissions data from M100 vehicles are sparse, they appear to be less variable (Lorang, 1989).

Raw NO_x emissions from methanol flexible-fuelled vehicles (using M85) can be lower than for petrol, although by increasing the compression ratio to take account of the increased octane rating of the fuel (as with M100), NO_x emissions can rise and may be higher than those from petrol, unless the air/methanol mixture is controlled at the stoichiometric ratio (6.4:1) in which case a three-way catalyst may be employed. However, highest efficiency is achieved using a lean air/fuel mixture and this may lead to a conflict between maximising fuel efficiency and minimising NO_x. Increasing the air/fuel ratio (operating lean) would lower *engine-out* NO_x levels but preclude the use of a three-way catalyst, potentially increasing the *controlled* levels of these emissions.

Output of CO appears to be variable, depending on whether the vehicle is set to run lean which leads to lower engine-out CO but more difficult cold starting, which could itself lead to higher CO emissions during this period. At the stoichiometric air/fuel mixture with a three-way catalyst the levels are comparable with, or slightly higher than, petrol (DeLuchi *et al*, 1988a). Overall CO emissions from methanol-fuelled vehicles are not considered to be significantly different from those fuelled by petrol.

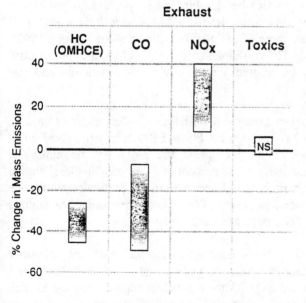

Figure 10. Effects of using M85 in prototype FFVs/VFVs

The US Auto/Oil Air Quality Improvement Research Program (AQIRP), outlined in section **2.2.3.1**, included a study of the effects of methanol (Gorse *et al*, 1992). Nineteen early (pre-1990) prototype flexible/variable fuel vehicles (FFV/VFV) were emission tested with industry standard petrol, an 85% methanol/petrol blend (M85), and a splash blend of M85 with M0 (petrol) giving 10% methanol (M10). Vehicle emissions were analysed for FTP exhaust emissions and evaporative emissions. The main effects of using M85 compared to M10 are shown in Figure 10 (from CRC, 1993).

The results showed reductions in HC (OMHCE – organic material hydrocarbon equivalent, does not reflect the weight of the oxygen atom in the total organic compound emissions whereas NMOG does) of about 37 percent and an average reduction in CO of 31 percent. NO_x emissions increased by about 23 percent when using M85. It is reported (CEC, 1993) that the NO_x emissions may not be representative of FFV/VFV and petrol vehicles that may be produced in the future. Since the prototypes were pre-1990 models, a comparison with the 1989 petrol fleet revealed NO_x emissions 18 percent lower with M85.

Evaporative emissions expressed as OMHCE were lower with M85 than with M0 but, as NMOG, were higher (both fuels had identical vapour pressures). Exhaust benzene, 1,3 butadiene and acetaldehyde emissions were lower in the FFV/VFV fleet with M85 compared with M0 by 84 percent, 93 percent and 70 percent respectively, but formaldehyde emissions were about five times higher. Aggregate exhaust toxics were 10 percent lower with M85, a value not considered as statistically significant.

3.6.1.2   Substituted for diesel

Where methanol is used as a diesel fuel substitute, the main emissions benefits arise from substantially lower particulates. CO emissions can be variable, depending on air/fuel ratio – used in a lean mixture (as with diesel) they are low and possibly lower than from diesel (Singh *et al*, 1987). NO_x emissions can also be variable, but proper design can ensure that methanol–fuelled diesel engines emit lower levels – MAN have built a methanol engine with NO_x emissions 50 percent those of the diesel–fuelled version (Cragg, 1992). The same comments with reference to HC emissions apply to diesel engines as to petrol – that is, higher overall levels may be emitted, but the majority is methanol and less photochemically reactive.

Automotive News (1991) has reported that the world's first heavy–duty engine to run on methanol or methanol–petrol blends has been United States EPA–certified. The Detroit Diesel Corporation's (DDC's) 6V–92TA engine meets the 1993 Federal emissions standards, achieves a thermal efficiency within 1–2 percent of a comparable diesel engine and produces less than half the NO_x emissions and one–fifth of the particulate emissions while performing as well as a diesel of equal power. The two–stroke diesel cycle engine, while certified for methanol (M100), can burn M85 and ethanol.

Table 7 overleaf (from Miller, 1991) shows emissions achieved with the methanol certification engine (with standard catalytic converter) compared with the 1993 Federal (1991 California) HD urban bus standard and with the equivalent engine using diesel fuel. The methanol engine's emissions are significantly lower than the equivalent diesel emissions – producing 60 percent of the HCs, 81 percent of the CO, 43 percent of the NO_x and one–fifth (19 percent) of the particulates. The table also reveals that using M85 instead of neat methanol would double CO emissions and significantly increase HCs, while the NO_x and particulates would remain low. It is also noteworthy that the diesel engine particulate emissions (highlighted in the table as *0.21* g/bhp–hr) are not within the 1991 California (1993 Federal) limits.

		Exhaust emissions g/bhp–hr				
		HC	CO	NO$_x$	Particulates	Formaldehyde
1991 California (1993 Federal) urban bus standard:		1.30	15.5	5.0	0.10	–
1991 6V–92TA diesel		0.50	1.60	4.6	*0.21*	–
6V–92TA methanol (% of diesel engine emissions in brackets)	**M100**	0.30 (60)	1.30 (81)	2.0 (43)	0.04 (19)	0.05
	M85	0.60 (120)	3.20 (200)	2.5 (54)	0.04 (19)	0.05

Table 7. DDC 6V–92TA methanol engine certified exhaust emissions

Bruetsch and Hellman (1992) report on the testing of a Volkswagen Jetta powered by a direct injection, neat methanol (M100) engine, based on a 1.9–litre turbocharged DI diesel engine. The engine uses glow plugs to assist in cold starting and light load operation, the compression ratio was lowered from 23:1 to 22:1 and exhaust gas recirculation was employed. Power output from the methanol engine was rated as 90 hp @ 4,500 rpm, compared with 68 hp @ 4,500 rpm from the turbodiesel.

A total of 22 FTP emission tests were carried out. The M100 Jetta, at its then (1992) state of development, emitted pollutants at levels below the California TLEV standards. Emissions of CO ranged from 0.1 to 0.5 g/mile, and averaged at below the levels from the diesel comparator vehicle. NO$_x$ emissions averaged 0.3 g/mile, considered to be very low for a lean–calibrated engine with oxidation catalyst. Non–methane HC emissions averaged less than 0.01 g/mile, with engine–out NMHC emissions lower than the tailpipe emissions of the conventionally–fuelled vehicle. Particulate emissions averaged 0.02 g/mile, well below those from the diesel vehicle.

Formaldehyde emissions averaged 4 or 5 mg/mile, reduced from 129 mg/mile engine–out average, indicating catalyst efficiency of about 96 percent for formaldehyde. The vehicle also demonstrated the ability to start at low temperatures (–29°C) and drive well. Fuel economy was slightly worse than the turbodiesel vehicle (but since power output was over 32 percent higher, a smaller capacity engine could be substituted).

3.6.1.3   Air toxics and secondary pollutants

Unburnt methanol accounts for about 90 percent of the HC emissions. While higher in absolute terms than from petrol, the reactivity of HC emissions from methanol is relatively low – estimated to be one–fifth that of petrol HC emissions (EPA, 1989). This explains why one of the key advantages quoted in the push for methanol as a vehicle fuel was the potential for lower ozone formation. Further experience from the United States suggests that FFVs using M85 yield around 30–40 percent net reduction in hydrocarbon–equivalent (adjusted for lower reactivity), or *effective* emissions (Gorse *et al*, 1992). Estimates of the hydrocarbon–equivalent emissions performance of dedicated vehicles burning pure methanol (M100) range between a 67 and 80 percent reduction relative to petrol–fuelled vehicles (Kohl, 1990).

Despite the expectation of lower hydrocarbon–equivalent emissions from methanol–fuelled vehicles leading to substantial reductions in ozone (O_3) formation, research in the United States has shown (Chang *et al*, 1989) that even with significant replacement of conventional–fuelled vehicles by methanol–fuelled ones, the reduction of O_3 levels appears to be modest – in the order of a few percent.

Other research has suggested (Holman *et al*, 1991, quoting Sierra Research Incorporated, 1989) that methanol vehicles could lead to more ozone formation than petrol vehicles, especially as the US Environmental Protection Agency's forecasts of emissions from methanol–fuelled vehicles have been based on relatively low mileage prototypes. The basis for Sierra Research's conclusions are not known, but are thought to relate to the low mileage prototypes tested and the expectation that vehicular emissions performance deteriorates with age.

Methanol will reduce significantly (or nearly eliminate, for M100) emissions of some toxic substances, primarily benzene, 1,3–butadiene and PAH (US Congress, 1990). As an example, a dedicated methanol vehicle would emit between five and 15 percent of the benzene from a petrol–fuelled one (Mechanical Engineering, 1991). This reduction has been cited by the EPA (1989) as a critical benefit of the substitution of methanol as a vehicle fuel.

The biggest emission disadvantage of methanol is the high formation of formaldehyde, a toxic and possibly carcinogenic substance and of most concern when emitted in enclosed spaces such as garages and tunnels. Chapter 11 discusses the environmental and health effects of a range of exhaust emissions, including aldehydes. Whereas petrol engines generally emit formaldehyde at rates of less than 10 mg/mile, methanol vehicles produce formaldehyde at rates several times this, early FFVs averaging 106 mg/mile over the vehicle life (US Congress, 1990). Vehicles fuelled with M85 have exhibited formaldehyde emission rates of between 22 and 37 mg/mile, and other tests by the US EPA have shown much lower levels than these, although for relatively new cars.

Nagai *et al* (1991) have reported that more than 90 percent of formaldehyde emissions occur when an engine is "cold", so cold catalytic activity is an area requiring improvement. Automotive manufacturers have expressed concern that long–term catalytic control of formaldehyde, over a vehicle lifetime, may represent a serious challenge to the industry. The 1993–model FFVs being produced in the US (see section **3.8**) are the first to have any semblance of production line manufacture. The area of long–term formaldehyde emission control is a major concern that will be especially scrutinised. Research into the use of electrically–heated catalysts and air injection during the cold–start period has shown (Newkirk *et al*, 1992) that average formaldehyde emission levels from M85 FFVs of less than 3 mg/mile are achievable during the FTP test, although these were from new vehicles prior to durability testing.

The use of neat methanol (M100) is expected to reduce the environmental and health effects of formaldehyde emissions, according to the US EPA (1989). This is because the majority of ambient formaldehyde is not thought to be due to direct vehicle emissions but formed indirectly in the atmosphere through photochemical reactions involving reactive HCs. Indirect formaldehyde formation with M100 vehicles is expected to decrease relative to petrol–fuelled vehicles due to the relative decrease in reactive HCs emitted.

With neat methanol use, the decreased amount of indirect formaldehyde formed is expected to offset any increase in direct formaldehyde emissions (ibid). However, a problem exists in that direct exhaust emissions of higher levels of formaldehyde with M100 may cause increased nuisance and health effects in populations living close to dense traffic activity, even though average ambient formaldehyde levels may be lower.

3.6.2 Life cycle emissions

When a fuel life cycle analysis is carried out for methanol, taking into account all the energy losses in the conversion of methane to methanol and subsequent distribution, in addition to the CO_2 generated during combustion (point of use), a range of CO_2 (and other) emissions are obtained relative to petrol. Several studies have estimated these greenhouse gas (GHG) emissions as CO_2 equivalent, and are shown in Figure 23 in chapter 10. The IEA (1990) quoting DeLuchi (1988b) show that methanol from natural gas demonstrates a 3 percent advantage over petrol, and were the methanol to be produced from coal, it would generate between 52 and 98 percent more CO_2 than petrol per vehicle mile driven.

Other studies show a range of estimates of greenhouse gas emissions from 12 percent better than from petrol, to 21 percent worse, using methanol produced from natural gas. However, the consensus seems to be that methanol produced from natural gas displays a fuel cycle GHG emissions advantage of up to 10 percent over the use of petrol as an automotive fuel. All conclude that methanol produced from coal would drastically add to greenhouse gas emissions – perhaps twice as much as from the use of petrol.

However, improved future technology such as methane recovery at the coal mine and CO_2 recovery at the production plant (both very expensive) could reduce the global warming impact to less than that from petrol from crude oil (EPA, 1989). If the long–term use of coal as an energy supply is to be considered, then these technologies need to be further developed.

3.7   Costs

3.7.1   Fuel production and distribution

Figure 26 in chapter 10 shows the IEA estimates of alternative fuel overall costs, relative to a baseline for petrol from crude oil, based on 1987 economics and fuel refining or feedstock conversion processes. Only at the low cost estimate of methanol from natural gas (111 percent of the petrol cost) does it appear that this fuel may be economically viable. However, as stressed in section **10.6**, improvements in the fuel conversion efficiency of perhaps up to 50 percent would help ensure methanol's economic viability as a fuel, as would significant crude oil price increases relative to the price of natural gas (used as the methanol feedstock). Santini (1988) asserts that most leading forecasters project an acceleration of crude oil prices in the 1990s and that the use of methanol produced from natural gas could reduce the costs and dependence of imported oil, but that the time to develop such a substitute is now while oil prices are still steady.

A Financial Times Management Report by Cragg (1992), from a review of various economic studies, estimates that methanol is 30–40 percent more expensive than petrol and 50–70 percent more expensive than diesel, on an equivalent energy basis. In late 1991 methanol was $125/tonne and premium leaded petrol $215/tonne. For the same energy supplied, the 1991 price for methanol would be the petrol equivalent of $270/tonne.

The United States EPA (1989) has estimated that the petrol–equivalent price for methanol, based on vehicle efficiency improvements of 5 percent above petrol, was competitive with the then–current US petrol prices, and that the dedicated methanol vehicle equivalent price would be around 20 percent cheaper, using (largely imported) natural gas as the feedstock. The US Department of Energy (DOE) (1991a) is sponsoring a research programme to reduce the cost of biomass–derived methanol from $0.75/gallon to $0.45/gallon, thereby making it competitive with that derived from natural gas. The research is focused on new and improved catalysts for the methanol conversion process.

3.7.2   Vehicle modification

The cost of vehicle modification to run on methanol or a petrol–methanol blend (such as M85) is somewhat variable, depending on whether the vehicle is modified or retro–fitted to accommodate the fuel, or if it specifically designed for it. Dual–fuel vehicles also require additional components, such as an alcohol sensor, to adjust the engine timing and fuel injection, depending on the alcohol concentration. Vehicles that use neat methanol (M100) as a fuel would probably require cold starting aids, such as fuel pre-heating, propane-assistance or direct fuel injection, again adding to the vehicle cost.

Estimates of the vehicle on-cost when using methanol as a fuel are most frequently obtained from American studies, since methanol fuel programmes have been underway in that country for many years. The US EPA (1989) have presented cost estimates provided by the Ford Motor Company for FFVs able to use M85 as a fuel. For a vehicle produced in high volume (100,000+ per annum), the incremental cost over that for a petrol-fuelled vehicle is $200 to $400. The California Air Resources Board (CARB) estimate the added costs for methanol-fuelled light-duty vehicles to be in the range $200 to $370 in order to comply with increasingly stringent emissions legislation (from the California TLEV through to ULEV standards).

A World Bank Technical Paper on alternative transport fuels estimates conversion costs for a range of alternatives, including methanol (Moreno and Fallen Bailey, 1989). Conversion costs for the following range of vehicles in 1989 were:

- Passenger cars $350
- Light duty trucks $500
- Heavy duty goods vehicles $3,200
- Buses $4,300

Limited production prototype FFVs that have been demonstrated in the US have cost significantly more than their petrol-fuelled counterparts. The US General Accounting Office (GAO, 1991) has summarised the incremental costs of procuring medium-sized alternatively-fuelled vehicles (65 methanol FFVs) in 1990. The average cost of the alternative fuel components was $2,250 and represents about 30 percent of the typical price of a government-procured compact petrol vehicle in 1990 ($7,730), or about 16% of the total price of the alternatively-fuelled vehicles themselves (at $14,130 each).

Ford and General Motors have provided testimony to the US Congress stating that for dedicated methanol vehicles produced in volumes of at least 100,000 per annum, the cost would be no more than that of a comparable petrol vehicle (EPA, 1989). The extra costs of more sophisticated fuel injection equipment (to help cold starting), methanol-compatible components, larger fuel tank and formaldehyde emission control components are expected to be offset by the use of a smaller, lighter engine (and compound weight saving effects), smaller cooling capacity and reduced evaporative emission controls.

3.8   Demonstration

The California Air Resources Board (CARB), the US Environmental Protection Agency (EPA), some individual fleet owners and most US automotive manufacturers have been operating methanol-fuelled vehicles for many years - up to a decade in some cases (US

Congress, 1992). Most automotive manufacturers worldwide have introduced alcohol vehicle programmes to develop engine technology should they decide (or need) to introduce vehicles able to use alcohol fuels.

In the US, until recently, most alternatively-fuelled vehicles have been prototypes or retrofitted under experimental permits. From 1992 Chrysler, Ford and General Motors have produced FFVs on production lines, mostly for supply to State and Federal government fleets (US Congress, 1992). Chrysler plans to produce at least 2,000 1993-model flexible fuel passenger cars and would expand production if sales demand grows. Ford has delivered over 1,100 methanol-capable cars since 1981, mostly to California, and has agreed to deliver 2,500 1993-model flexible fuel cars, certified to California's 1994 TLEV standards. General Motors has agreed to deliver up to 4,200 flexible fuel passenger cars in 1992 and 1993, also to California. The supply of these vehicles is aided by the Federal Government's Alternative Motor Fuels Act of 1988 which mandated the procurement of alternatively-fuelled government vehicles and provides assistance for truck and bus alcohol (and natural gas) application and demonstration projects (US DOE, 1991b).

Heavy-duty diesel engine methanol programmes have also been undertaken by many engine manufacturers with encouraging particulate and NO_x emission results (see section **3.6.1.2**). The Detroit Diesel Corporation's (DDC's) 6V-92TA engine that meets the 1993 Federal emissions standards is shown in Figure 11 (from Miller, 1991). Engine modifications include new thicker pistons in order to provide a higher compression ratio, the turbocharger and control system was modified to allow greater airflow, higher flow fuel injectors were necessary for the greater fuel delivery volume and a glow plug system was developed that is used for starting and to assist combustion during warmup.

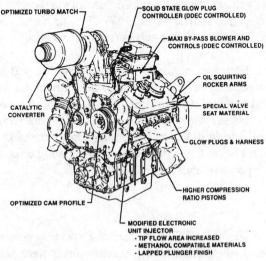

Materials changes include stainless steel fuel system components and powdered stainless steel for the manufacture of the (usually aluminium) fuel injectors. A close-coupled oxidation catalytic converter using platinum has been included to reduce emission levels even further.

Figure 11.   DDC methanol bus engine

ICI, in conjunction with the Los Angeles South Coast Air Quality Management District, launched a trial in 1989 involving two diesel-engined buses using methanol and ICI's ignition improver, Avocet. This trial was extended in 1991 to include 12 buses to gain more comprehensive emissions data (Professional Engineering, 1991).

The Japanese and European manufacturers have not announced equivalent large-scale production plans, but have nevertheless undertaken extensive research and development and provided several prototypes to US agencies, such as CARB and the EPA. Volkswagen, for example, introduced a 300-strong passenger car fleet to California in 1990 (Volkswagen AG, undated). Automotive Engineering (1991) reports that Nissan is close to introducing a FFV to the market, by providing a fleet of flexible fuel passenger cars (NX coupe) to the California Energy Commission (CEC) for testing and evaluation. The vehicles employ an electrostatic fuel sensor to determine the alcohol concentration instead of the more common TNO-developed optical fuel sensor. Volvo of Sweden has undertaken light and heavy-duty engine methanol research and development and has a fleet of FFVs on test in California, and is experimenting with electrically heated catalysts to reduce formaldehyde emissions (Mechanical Engineering, 1991).

Of the West European countries, Germany has carried out most R&D and implementation with methanol, partly in response to the first energy crisis in 1973. Automotive News (1992) announced that Mercedes-Benz plans to deliver six flexible fuel vehicles to the CEC during 1992 and to produce 9,000 passenger cars for sale in California in 1994. Other European motor manufacturers with a high profile alcohol fuels programme include Saab who have built a prototype 9000 model capable of using methanol, ethanol and petrol, in any ratio (Autocar and Motor, 1992). Peugeot-Citroën (PSA) has had experience with methanol as a fuel since the 1970s and has announced new research beginning in 1990.

Methanol has been considered as an octane booster in some Asian countries but not as a primary petrol substitute, except in China which has large coal reserves used as the feedstock (Sathaye *et al*, 1988). In Japan, the Ministry of Transport (MOT) and the Ministry of International Trade and Industry (MITI) have been preparing strategies for methanol introduction, primarily for M85 to M100. A research project was started in 1984 to verify the feasibility of using methanol and in 1989 a 12-vehicle test fleet, running on M85, began a three-year trial evaluation (Hoshino and Iwai, 1991). The project's aims are to accumulate data relating to vehicle driveability, durability, exhaust emissions and fuel economy in realistic driving conditions. It is not a project aim to concentrate on the practical use of methanol as a vehicle fuel in the near future.

3.9 Outlook

It has been demonstrated that methanol can be used as a vehicular fuel, albeit in mostly modified vehicles to date, although flexible fuel vehicles are now being mass produced in the United States. Dedicated vehicles able fully to realise the benefits methanol has to offer are still at the experimental stage. Methanol is amongst the most "ready" of the alternative fuels because of extensive experience of use of the fuel. Certain claimed emissions advantages seem unclear, and there is no clear evidence that its use would reduce transport's contribution to global warming. Perhaps the most promising application for methanol is as a fuel for heavy–duty diesel engines where the reductions in NO_x and especially particulates are beneficial for urban fleet operations.

The difficulties in providing an infrastructure and the uncertain economics of methanol as a vehicle fuel, especially in the early stages of its introduction (when economies of scale cannot be achieved), imply that its widespread use is unlikely to progress without government promotion (such as provided by the State of California) or substantial and lasting increases in oil prices. Additionally, improvements in the fuel economy and emissions control of conventional vehicles and the indications that refiners can restructure the composition of conventional fuels to help reduce emissions, imply that a further tightening of vehicle pollution standards may be possible. This may remove some of the pressure for a switch to certain alternative fuels, such as methanol (US Congress, 1990).

However, since the most likely feedstock for large–scale methanol production would be natural gas (methane), the question arises whether it is better to convert natural gas to a liquid fuel, or use it directly (stored in compressed or liquefied form) as a gaseous fuel. Methane (natural gas) as a vehicle fuel displays certain benefits and disadvantages, as discussed in chapter 7. The conversion of natural gas to methanol uses more energy than that to compress (or liquefy) methane to use as a fuel itself, but methanol provides better vehicle operating range for a given size of fuel storage system. A country's natural gas infrastructure may also play a part in determining methanol's feasibility – with an established gas distribution network it may be cheaper and less energy intensive to use gas as a fuel rather than transport liquid by tanker, whereas methanol is cheaper to transport over long distances than gas (both by tanker) in the absence of such a network.

3.10   Summary

Methanol is a liquid fuel with a high octane rating (RON=111) but a lower energy density than petrol or diesel (requiring a volume of 1.8, falling to perhaps 1.5 times that of petrol for equivalent vehicle range). Methanol has a low cetane rating, thereby requiring spark or ignition assistance or a fuel additive in a diesel engine.

<u>As a petrol substitute</u>

↓ evap. emissions, ↓ HC reactivity (by about ⅔), CO variable but no significant difference.

NO_x emissions variable: ↑ with increased compression ratio (dedicated engine), may ↓ in FFVs. Using λ=1 with 3WC will produce NO_x emissions comparable with petrol.

much ↑ formaldehyde, although much ↓ benzene, ↓ 1,3–butadiene.

Used in FFVs (as M85), HC emissions are variable, but considered less reactive overall (by about ⅓) than from petrol.

<u>As a diesel substitute</u>

substantially ↓ particulates (80% lower) and much ↓ NO_x (50% lower).

↓ HC, ↓ CO.

↓ PAH.

REFERENCES FOR CHAPTER 3

AUTOCAR AND MOTOR (1992). Dispatches: Saab has built a prototype 9000 .... Page 11, 1 January 1992.

AUTOMOTIVE ENGINEERING (1991). Tech briefs: Nissan methanol–gas vehicle. Pages 58–59, Volume 99, Number 6, June 1991.

AUTOMOTIVE NEWS (1991). Detroit Diesel certifies alternative–fuelled engine. 5 August 1991.

AUTOMOTIVE NEWS (1992). Big 3 prepare methanol cars for showrooms (page 1). Fuels: Methanol cars debut this year (page 30). 6 January 1992.

BABER B B, S J LESTZ and M E LEPERA (1990). Technology demonstration of US Army methanol–fuelled administrative vehicles. SAE Technical Paper Series No 902158, Society of Automotive Engineers, Inc., Warrendale, Pennsylvania, United States of America.

BRUETSCH R I and K H HELLMAN (1992). Evaluation of a Passenger Car Equipped with a Direct Injection Neat Methanol Engine. SAE Technical Paper Series No 920196, Society of Automotive Engineers, Inc., Warrendale, Pennsylvania, United States of America.

CEDIGAZ (1989). Natural gas in the world in 1988. Cedigaz (France), Paris.

CHANG T Y, S J RUDY, G KUNTASAL and R A GORSE Jr (1989). Impact of methanol vehicles on ozone air quality. Atmos Environ, Volume 23, Number 8, Pages 1629–1644.

COORDINATING RESEARCH COUNCIL (1993). Auto/Oil Air Quality Improvement Research Program – Phase I Final Report. May 1993. CRC Inc., Atlanta, Georgia, United States of America.

CRAGG C (1992). Cleaning up motor car pollution: New fuels and technology. Financial Times Management Report, FT Business Information Ltd., London.

DELUCHI M A, R A JOHNSTON and D SPERLING (1988a). Methanol vs. natural gas vehicles: A comparison of resource supply, performance, emissions, fuel storage, safety, costs, and transitions. SAE Technical Paper Series No 881656, Society of Automotive Engineers, Inc., Warrendale, Pennsylvania, United States of America.

DELUCHI M A, R A JOHNSTON and D SPERLING (1988b). Transportation fuels and the greenhouse effect. Transportation Research Record 1175, National Research Council, Washington, DC, United States of America.

GARDINER D P, V K RAO, M F BARDON and V BATTISTA (1991). Improving the cold start combustion in methanol fuelled spark ignition engines by means of prompt EGR. SAE Technical Paper Series No 910377, Society of Automotive Engineers, Inc., Warrendale, Pennsylvania, United States of America.

GORSE Jr R A, J D BENSON, V R BURNS, A M HOCHHAUSER, W J KÖHL, L J PAINTER, R M REUTER, B H RIPPON and J A RUTHERFORD (1992). The Effects of Methanol/Gasoline Blends on Automobile Emissions. SAE Technical Paper Series No 920327, Society of Automotive Engineers, Inc., Warrendale, Pennsylvania, United States of America.

HOLMAN C, M FERGUSSON and C MITCHELL (1991). Road transport and air pollution: Future prospects. Rees Jeffreys Discussion Paper 25, Transport Studies Unit, Oxford University.

HOSHINO T and N IWAI (1991). A demonstration of prototype near–neat methanol vehicle controlled fleet test. Int J of Vehicle Design, Volume 12, Number 2, Pages 229–239.

INTERNATIONAL ENERGY AGENCY (1990). Substitute fuels for road transport: A technology assessment. OECD/IEA, Paris.

JOHNS R A, A W E HENHAM and S K C NEWNHAM (1991). The combustion of alcohol fuels in a stationary spark–assisted diesel engine. Proceedings of the IMechE conference "Internal combustion engine research in universities, polytechnics and colleges", 30–31 January 1991. Mechanical Engineering Publications Ltd.

KOHL W L (1990). Methanol as an alternative fuel choice: An assessment. Foreign Policy Institute, The Johns Hopkins University, Washington, DC, United States of America.

LEE C I, E S KIM and K K CHO (1987). Performance and exhaust emission in spark ignition engine fuelled with methanol–butane mixture. Proceedings of the 4th International Pacific Conference on Automotive Engineering "Mobility: the technical challenge", 8–14 November 1987, Melbourne, Australia.

LORANG P A (1989). Emissions from gasoline–fuelled and methanol vehicles. Proceedings from "Conference on Methanol as an Alternative Fuel Choice: An Assessment", 4–5 December 1989, Johns Hopkins Foreign Policy Institute, Washington, DC, United States of America.

MARTIN D and L MICHAELIS (1992). Road transport and the environment: Policy, technology and market forces. Financial Times Management Report, FT Business Information Ltd., London

MECHANICAL ENGINEERING (1991). Alternative fuels: Paving the way to energy independence. Pages 42–46, December 1991.

MILLER S P (1991). DDC's production 6V–92TA methanol bus engine. SAE Technical Paper Series No 911631, Society of Automotive Engineers, Inc., Warrendale, Pennsylvania, United States of America.

MORENO R Jr and D G FALLEN BAILEY (1989). Alternative Transport Fuels from Natural Gas. World Bank Technical Paper number 98. Industry and Energy Series. Washington, DC, United States of America.

NAGAI K, T MORI, F TSURUGA and N IWAI (1991). The behaviour of hydrocarbon and oxygen containing compounds emitted from otto–type methanol–fuelled (M85) vehicles. JSAE Rev, Technical Notes, pages 64–66, July 1991.

NEWKIRK M S, L R SMITH, M AHUJA, S ALBU, S SANTORO and J LEONARD (1992). Formaldehyde emission control technology for methanol–fuelled vehicles: Catalyst selection. SAE Technical Paper Series No 920092, Society of Automotive Engineers, Inc., Warrendale, Pennsylvania, United States of America.

PROFESSIONAL ENGINEERING (1991). Transport fuels: Making motors change their tipple. Pages 30–31, April 1991.

SAKAI T, I YAMAGUCHI, M ASANO, T AYUSAWA and Y K KIM (1987). Transient performance development on dissociated methanol fuelled passenger car. Proceedings of the 4th International Pacific Conference on Automotive Engineering "Mobility: the technical challenge", 8–14 November 1987, Melbourne, Australia.

SANTINI D J (1988). The past and future of the petroleum problem: The increasing need to develop alternative transportation fuels. Center for Transportation Research, Argonne National Laboratory, Argonne, Illinois, United States of America.

SATHAYE J, B ATKINSON and S MEYERS (1988). Alternative fuels assessment: The international experience. International Energy Studies Group, Lawrence Berkeley Laboratory, University of California, Berkeley, California, United States of America.

SEPPEN J J and R R J TER RELE (1989). Gaseous and methanol fuels in lean burn and stoichiometric engine concepts with emphasis on exhaust gas emissions – thermodynamic simulations. Proceedings of the 2nd International Conference on New Developments in Powertrain and Chassis Engineering, Palais du Congress, Strasbourg, 14–16 June 1989. Mechanical Engineering Publications Limited.

SIERRA RESEARCH INCORPORATED (1989). Potential emissions and air quality effects of alternative fuels – final report. SRI, Sacramento, California, United States of America.

SINGH M K, C L SARICKS, S J LABELLE and D O MOSES (1987). Emerging environmental constraints on the use of gasoline and diesel fuel and tradeoffs associated with the use of alternative fuels. Center for Transportation Research, Argonne National Laboratory, United States of America.

US CONGRESS (1990). Replacing gasoline: Alternative fuels for light–duty vehicles. Office of Technology Assessment, US Government Printing Office, Washington, DC, United States of America.

US CONGRESS (1992). Alternative fuels for automobiles: Are they cleaner than gasoline? CRS Report for Congress 92–235 S, Congressional Research Service, Washington, DC, United States of America.

US DEPARTMENT OF ENERGY (1991a). Technology Factsheet No 10.0: Biofuels research. US DOE, Office of Transportation Technologies, Washington, DC, United States of America.

US DEPARTMENT OF ENERGY (1991b). Technology Factsheet No 6.0: Alternative Motor Fuels Act demonstration projects. US DOE, Office of Transportation Technologies, Washington, DC, United States of America.

US ENVIRONMENTAL PROTECTION AGENCY (1989). Analysis of the economic and environmental effects of methanol as an automotive fuel. Special Report, Office of Mobile Sources, EPA, Washington, DC, United States of America.

US GENERAL ACCOUNTING OFFICE (1991). Alternative fuels: Increasing Federal procurement of alternative–fuelled vehicles. US GAO, Report to Congressional Requesters, GAO/RCED–91–169, Washington, DC, United States of America.

VOLKSWAGEN AG (Undated). VW Documentation: Research for the Future: Alternative Fuels. Research and Development/Public Relations, Volkswagen AG, Wolfsburg, Germany.

WATERS M H L (1992). Road vehicle fuel economy. TRRL state–of–the–art review 3. HMSO, London.

ZELENKA P, P KAPUS and L A MIKULIC (1991). Development and optimisation of methanol fuelled compression ignition engines for passenger cars and light duty trucks. SAE Technical Paper Series No 910851, Society of Automotive Engineers, Inc., Warrendale, Pennsylvania, United States of America.

4.   ETHANOL

4.1   Introduction

Ethanol is the most widely used alternative vehicle fuel in the world. Brazil is the largest producer and consumer of ethanol with two and a half million dedicated vehicles using the fuel in 1986 rising to over four million in 1990, requiring 11–12 billion (10^9) litres of ethanol a year that is estimated to replace 200,000 barrels of oil per day (Cragg, 1992). The United States produced around three billion litres in 1986 which compares with only 1.8 *million* litres of ethanol produced by the whole of Europe in that year (Holman *et al*, 1991). In 1992 the United States ethanol production capacity was expected to be 4.6 billion litres and there are plans to increase this to almost seven billion litres (Hall *et al*, 1992).

Ethanol possesses similar properties and combustion characteristics to methanol and can be used as a fuel by itself or otherwise blended with petrol in varying amounts. Ten percent ethanol mixed with petrol has been sold as "gasohol" in the United States since 1979 whereas in Brazil an 85/15 to 95/5 percent ethanol/petrol mix is commonly used.

Its practical use is very similar to that of methanol – it is a liquid at room temperature and so can be stored in conventional fuel tanks and can be used in conventional engines (spark and compression ignition with assistance) with relatively few modifications. While ethanol possesses a lower energy density than petrol, it is higher than that of methanol and therefore requires a smaller quantity for a given vehicle operating range. Ethanol or ethyl alcohol is also known as drinking alcohol and consequently does not suffer, to the same extent, from some of methanol's safety concerns relating to toxicity.

Although methanol is acknowledged generally as the least expensive of the alcohol fuels, ethanol has gained certain support. Proponents of ethanol usage either as a blending agent or a neat fuel argue that its expanded use as an automotive fuel will displace (imported) oil, aid the agricultural economy by creating a stable market for its farmed feedstocks, and improve air quality by reducing emissions from vehicles using it (US Congress, 1990).

4.2   Fuel characteristics

As with methanol, ethanol's properties are suited to improving the efficiency of the spark–ignition engine. Section 3.2 discusses these aspects due to methanol as the fuel, most of which apply equally to ethanol. The engine efficiency and power enhancements (over those from petrol) are generally less with ethanol than for methanol (Waters, 1992), partly due to the slightly lower octane rating of the fuel – RON 108 (Zelenka *et al*, 1991).

Ethanol, like other alcohols, possesses a low cetane rating (8, compared with 5 for methanol, and typically 50 for diesel fuel), making it unsuitable for compression ignition without an ignition enhancer or initiator. While the mass energy density of ethanol is 35 percent higher than that of methanol (27 MJ/kg compared with 20 MJ/kg), it is still low with respect to petrol or diesel, calling on increased volume to provide identical vehicle range. About 50 percent additional ethanol volume is currently required, but with the development of dedicated engines, the volume of ethanol is expected to fall to 1¼ times that of petrol.

Some of the limitations of neat ethanol, such as cold–start problems, can be overcome by blending ethanol with petrol to from E85 which is commonly used in flexible fuel vehicles (FFVs), but this tends to diminish the inherent benefits of using ethanol in the first place, such as its low volatility and lack of benzene. The evaporative emissions of neat ethanol (E100) are lower than from petrol, but higher than diesel fuel. E85 evaporative emissions are expected to be comparable with those from petrol, although in practice FFVs running on mixtures of E85 are likely to have higher evaporative emissions than petrol vehicles (Pitstick, 1993). This is because of increased fuel tank size and/or more frequent refuelling because of the lower energy density of the fuel.

4.2.1   Safety implications

The corrosive properties of methanol apply largely to ethanol also, as does the low flame luminosity and increased likelihood of vapour ignition in enclosed spaces (lower and upper flammability limits of 3.3% and 19.0%, compared with 1.4% and 7.6% for petrol by volume, in air). While ethanol is toxic to marine life in high concentrations (following a spill, for example) it is readily biodegradable and will evaporate quickly if spilled on land (US EPA, 1989). Also, contamination of water supplies is less dangerous than for methanol or petrol because ethanol is less toxic to humans in equal concentrations and has a recognizable taste.

4.3   Feedstocks

Ethanol can be produced from biomass (sugar cane, sugar beet, wood, corn and other grain) and natural gas or crude oil. The fossil fuel route involves the catalytic hydration of ethylene (the feedstock), itself produced from methane or oil. The production of ethanol from biomass sources involves the fermentation and distillation of the crop, in a similar process to that producing alcohol for human consumption.

In Brazil sugar cane is the prime feedstock source while corn provides most of the United States' ethanol. Researchers are examining the possibility of using other plant matter for ethanol production – in fact virtually any starch producing plant can be turned into alcohol. Research funded by the US Department of Energy has shown that woody and herbaceous crops have high potential as feedstocks for the production of biofuels such as ethanol. The Biofuels Feedstock Development Program (BFDP) has identified fast–growing trees with high energy potential and various herbaceous crops including grasses as being particularly suitable, and with anticipated future technology improvements the same crop area will supply over twice the fuel that can be obtained with current production and conversion processes (Wright, *undated*).

Bioethanol effectively needs a high sugar content crop because the fermentation process generally fails to convert all the existing sugar to alcohol. The French attempted to use the Jerusalem artichoke as a source of sugar in the early 1980s but found the best combination to be sugar beet plus wheat. Their project was to meet a seven percent ethanol/petrol blend and was calculated to require 1½ million litres using 150,000 hectares of sugar beet and 350,000 hectares of wheat, or 5,100 square kilometres in total. Germany proposed a similar project using 200,000 hectares of land to produce 250,000 tonnes of ethanol from 1¼ million tonnes of grain.

Sweden, looking at a six percent ethanol/petrol blend, believed that it could be met with 50 percent of the country's 1-2 million tonnes of annual production of surplus wheat. The UK Department of Energy has calculated that, in theory, there was sufficient excess wheat and barley to substitute ethanol for five percent of the petrol demand. Generally, however, agricultural land in Europe is in shorter supply than in Brazil, and that ethanol as a fuel in Europe is seen more as a potential means to absorb surplus agricultural production, by using it in low level blends as an octane enhancer (ibid).

A pan-European consortium lead by the biotechnology centre at Imperial College has developed a process to turn poisonous paper-making waste into ethanol (The Engineer, 1992). Agricultural waste such as cereal straws may also be processed, with ethanol one of the end products. A special bacterium strain has been developed that results in very little unwanted by-products, with encouraging consequences for reduced production cost.

4.4   Infrastructure

If ethanol is to be produced from biomass, large areas of land are needed to grow the crop to produce the fuel. To supply Brazil's 11-12 billion litres of ethanol per year has required three million hectares (about 30,000 square kilometres or 12,000 square miles - 109 miles by 109 miles) to be planted, each hectare producing 50-65 tonnes of cane and each tonne of cane yielding around 70 litres of ethanol (Cragg, 1992). However, productivity increases of over 43 percent during the eight year period 1977-1985 in bioethanol production in Brazil (up from 2,663 litres/hectare to 3,811 l/ha) give cause for continued expectations of productivity increases and reduced land area requirements in the future (de Oliveira, 1991).

For every litre of alcohol produced, there are 12-13 litres of waste (high biochemical oxygen demand stillage) which is theoretically recycled as fertiliser. The market for ethanol distillation byproducts has a fundamental impact on the cost of producing the fuel (US Congress, 1990). If the stillage can be sold as a high-protein substitute the net feedstock costs fall substantially. If markets were to become saturated, the stillage would have a much lower value and could even represent an additional cost for disposal. In this case additional stillage distribution and disposal infrastructure would be required.

Not least of the problems associated with biofuels is the fact that agriculture is cyclical. In Brazil, 80 percent of the total ethanol production takes place between late May and early September. Consequently, this production has to be stored and in effect the available tankage must be capable of holding up to two–thirds of annual demand. Brazil currently requires 10 billion litres, or more, of storage. Additionally the quality of the crop can be substantially affected by the weather – an early Brazilian harvest can reduce the effectiveness of sugar cane as an alcohol feedstock by as much as 30 percent.

Many of the issues relating to the distribution infrastructure required for the implementation of methanol as a vehicle fuel apply equally to ethanol. There is no need, therefore, to repeat these – chapter 3 discusses the implications for methanol and the rest of this section will simply highlight the *differences*, where they exist, between the two fuels. The remainder of this chapter will follow this format, discussing the differences between ethanol and methanol in terms of each specific section heading and the issues raised therein.

A greater volume of ethanol would need to be distributed for the equivalent petrol it replaces, due to its lower energy density. However, the extra volume required is about 20 percent less than that of methanol – 1½, falling to possibly 1¼ times that of petrol when used in dedicated engines and vehicle designs. Ethanol/petrol mixtures have low tolerance to water at low ethanol concentrations (such as gasohol – 10% ethanol). A very small amount of water can cause phase separation and increase the rate of corrosion in the engine (Holman *et al*, 1991). Hence the distribution system requires measures to prevent water ingress, as with methanol.

4.5   Vehicle modifications

4.5.1   Petrol–engined vehicles

Many of the vehicle conversion changes and storage implications for methanol apply equally to ethanol, being an alcohol also. Spark–ignition engines require certain modifications including engine management changes and material substitution (some metal alloys and polymers) in the fuel system. To take advantage of ethanol's higher (than petrol) octane rating, the compression ratio should be increased, thereby improving thermal efficiency. Section **3.5.1** presents an overview of the vehicular implications for methanol as a petrol substitute, and is generally applicable to other alcohols, including ethanol.

4.5.2 Diesel–engined vehicles

Because of ethanol's low cetane value, its use in compression ignition (CI) engines requires some form of ignition enhancement, such as by the addition of ignition–improving agents and the other is through emulsification of the alcohol fuel in diesel fuel, but does not enable complete substitution for diesel fuel. Ignition–improving agents, such as ICI's "Avocet", provide a means of allowing diesel engines to operate on ethanol when the combustion air is not preheated. Once a diesel is started, it can operate on pure alcohols under high–speed, high–load conditions, but cold–start, part–load and transient operation is not possible due to misfiring as the engine temperature falls. Quite high levels of around 5 percent (or higher in cold climates) of costly ignition improvers have to be used for reliable alcohol–fuelled CI engine operation.

The ways in which the diesel engine can be modified and adapted to allow the use of straight alcohols in spite of their poor self–ignition properties include:

- A two–fuel system of partial replacement of the diesel fuel by alcohol carburation (or "fumigation") of the air inducted to the engine;

- Duplication of the injection system, one for diesel fuel as a pilot injection to initiate combustion and one for the ethanol as main fuel (dual–fuel); and

- The ignition of injected ethanol by spark plugs or glow plugs for the complete replacement of the diesel fuel.

The injection of two fuels into the diesel engine combustion chamber has proved suitable for ethanol if fuel pumps and injectors are externally lubricated or redesigned to compensate for the poor lubricity of alcohols. The system has the potential for very high replacement of the diesel fuel, from 75–90 percent depending on the duty cycle. A 100 percent replacement would be possible if the high–cetane pilot fuel consisted of ethanol with sufficient ignition improver to burn a number of fuels with low ignitability as secondary or main energy fuel (e.g. water–containing alcohols). The drawback of the dual–fuel, dual–injection concept is the increased cost of a second fuel system. Thermal efficiency has been shown to be as good as, or better than, the diesel counterpart.

4.6 Emissions performance

4.6.1 Vehicle exhaust emissions

4.6.1.1 Substituted for petrol

Ethanol has a lower carbon to hydrogen ratio than petrol or diesel fuel, and therefore produces less carbon dioxide when burnt. On an equivalent energy basis, ethanol produces 96 percent of the CO_2 emitted from the combustion of petrol (see Table 25 in chapter 10). Because the use of ethanol improves engine efficiency, on a per unit distance travelled basis, allowing for a 15 percent efficiency improvement, ethanol produces around 83 percent of the CO_2 of a petrol vehicle.

Ethanol–fuelled vehicles (E85 and higher) are expected to emit more ethanol and acet-aldehyde in the exhaust than petrol–fuelled vehicles, but total exhaust HC emissions should decrease. Experience in Brazil with uncontrolled engines using ethanol has shown aldehyde emissions to be five times higher than for petrol (Szwarc and Branco, 1985). NO_x formation can be lower with neat ethanol than with petrol, because the combustion temperature will be lower, but treated emissions are expected to be similar to petrol with comparable emission control equipment. Similarly, there is reportedly no reason to expect much lower CO emissions from neat ethanol vehicles (Pitstick, 1993).

Nevertheless, research at AVL in Austria, reported by Zelenka (1991), has shown generally lower exhaust emissions from an ethanol (E100)–fuelled 2.0-litre engine capacity passenger car than from its petrol–fuelled counterpart. During the US FTP 75 test cycle HC emissions were 0.1 g/mile (petrol 0.14 g/mile), CO 0.9 g/mile (petrol 0.8 g/mile) and NO_x emissions 0.2 g/mile (petrol 0.6 g/mile). Aldehyde emissions averaged at 14 mg/mile.

The lean combustion capability of ethanol could decrease CO emissions and might allow a US 1.0 g/mile NO_x standard to be met without a three–way catalyst, but future lower standards are expected to require stoichiometric fuelling (AFR of 9:1) with catalysts (ibid). Because the vapour pressure of ethanol is lower than that of petrol, lower evaporative losses can be expected.

4.6.1.2   Substituted for diesel

Where ethanol is used as a diesel fuel substitute, the main emissions benefits arise from substantially lower particulates. Alcohol–fuelled heavy–duty engines developed by Detroit Diesel Corporation (DDC) have demonstrated the ability to meet future EPA standards for HC, CO and particulates, as shown in Table 8 overleaf (from Pitstick, 1993). The engine tested was fitted with an oxidation catalyst, but while near–term NO_x limits are met, these emissions presently exceed the 1998 standard.

Research at the Helsinki University of Technology on an ethanol–fuelled spark–assisted 3.3–litre diesel engine fitted with an oxidising catalytic converter has shown very low emissions of HC and CO, with NO_x levels that satisfy the current ECE R49 standard. (Niemi and Ubong, 1991). Emissions are as follows (ECE R49 limits in brackets) – HC 0.24 g/kWh (3.5), CO 0.031 g/kWh (14) and NO_x 11 g/kWh (18).

Exhaust emission standard / emissions performance	Exhaust emissions g/bhp-hr			
	THC	CO	NO$_x$	Particulates
1993 Federal urban bus ('94 Heavy–duty)	1.30	15.5	5.0	0.10
1994 Federal urban bus	1.30	15.5	5.0	0.05
1998 Federal heavy–duty	1.30	15.5	4.0	0.10
1998 Federal urban bus	1.30	15.5	4.0	0.05
1991 6V–92TA diesel engine emissions	0.50	1.60	4.6	0.21
6V–92TA ethanol (E95) engine emissions	0.73*	1.71	4.15	0.04

* organic material hydrocarbon equivalent (OMHCE)

Table 8.   DDC 6V–92TA ethanol–fuelled engine certified exhaust emissions

4.6.1.3   Air toxics and secondary pollutants

Ethanol will reduce significantly (or nearly eliminate, for E100) emissions of some toxic substances, primarily benzene, 1,3–butadiene and PAH.   The biggest emission disadvantage of ethanol is the high formation of acetaldehyde, although it is less toxic than formaldehyde (Pitstick, 1993).   However, ethanol reacts in the atmosphere to create more acetaldehyde which itself reacts to form peroxyacetyl nitrate (PAN), which plays a role in ozone formation and is toxic.   Formaldehyde emissions from the use of ethanol are expected to be about the same as or slightly higher than from petrol engines.   The application of three–way catalyst technology effectively reduces ethanol and aldehyde exhaust emissions to an acceptable level (ibid).

Zelenka (1991) reports at development efforts directed towards improving the warm–up time of alcohol–fuelled engines, such as by employing heating elements in the inlet manifold.   Because most aldehyde emissions are generated when the engine is cold, effective pre–heating of the fuel mixture is seen as one important method of controlling these emissions.   Chapter 11 discusses the environmental and health effects of a range of exhaust emissions, including aldehydes.

Limited ozone modelling of speciated emissions from neat ethanol–fuelled vehicles suggest that the ozone–forming potential of ethanol is less than that of petrol and diesel, about the same as that of reformulated petrol, and higher than that of other alternative fuels (methanol, LPG, methane) (Pitstick, 1993).

4.6.2   Life cycle emissions

Fuel life cycle emissions of greenhouse gases (GHGs), usually expressed in CO_2 equivalent, are not well established for ethanol, due to the wide range of suitable feedstocks used and uncertainties of all the energy inputs into the whole cycle. Chapter 10 discusses various estimates of GHG emissions from alternative fuels. A study of the effect of ethanol produced from US corn crops revealed a GHG emission increase of between 15 and 36 percent, relative to petrol, depending upon various assumptions made.

More recent research reported by Pitstick (1993), quoting DeLuchi (1991), discusses what are considered to be the most comprehensive estimates of total GHG emissions from ethanol. The best estimate is that ethanol from wood (cellulosic biomass) reduces the per–mile emissions by about 70 percent. The results for ethanol from corn are not as promising (as shown in **10.3.2**) with an expected increase in GHG emissions by about 20 percent for light–duty vehicles, compared with petrol, and an increase of up to 50 percent for heavy–duty vehicles, compared with diesel fuel. Changes in assumptions, however, have a significant effect on the estimates, and arguments of the actual impacts of nitrogen fertilisers, for example, continue.

4.7   Costs

4.7.1   Fuel production and distribution

Figure 26 in chapter 10 shows the IEA estimates of alternative fuel overall costs, relative to a baseline for petrol from crude oil, based on 1987 economics and fuel refining or feedstock conversion processes. Ethanol produced from biofuels are estimated at between 2½ to 3¾ times the cost of petrol at 1987 fuel prices and level of technology. However, as stressed in section **10.6**, improvements in the fuel conversion efficiency in both the short term and in the long term (beyond 2000, ethanol's prospects may improve given the potential for relatively inexpensive production from wood and wastes, based on the enzymatic hydrolysis process currently being actively developed) will help to reduce the cost disadvantage to a degree.

A Financial Times Management Report by Cragg (1992), from a review of various economic studies, estimates that ethanol would not be competitive with petrol from crude oil, until crude prices rose to between $40 and $65/bbl, based on current ethanol production processes and costs. This is in broad agreement with BP Chemicals who believe that a $55 barrel of crude oil is necessary for current ethanol production to be cost–effective for motor fuel application (ENDS, 1992).

The US Department of Energy (DOE) (1991) is sponsoring a research programme to reduce the cost of biomass–derived ethanol from $3.60/gallon to $0.60/gallon by the year 2000. Over the last ten years the cost has fallen from $3.60 to $1.35. The research is focused on enzymatic hydrolysis, known as simultaneous saccharification–fermentation (SSF). Cellulosic feedstocks are treated with dilute sulphuric acid to solubilise the hemicellulose fraction, leaving solid cellulose and lignin. The cellulose fraction is exposed to cellulase enzyme together with yeast. The cellulose is decomposed to glucose, and yeast present in the SSF broth rapidly ferments the glucose to ethanol. This process reduces equipment costs and gives high yields and concentrations of ethanol, thereby reducing the fuel price (Josephson, 1992).

4.7.2 Vehicle modification

The cost of vehicle modification to run on ethanol or a petrol–ethanol blend (such as E85) is somewhat variable, depending on whether the vehicle is modified or retro–fitted to accommodate the fuel, or if it specifically designed for it. Dual–fuel or flexible fuel vehicles also require additional components, such as an alcohol sensor, to adjust the engine timing and fuel injection, depending on the alcohol concentration. Vehicles that use neat ethanol (E100) as a fuel would probably require cold starting aids, such as fuel pre–heating, propane–assistance or direct fuel injection, again adding to the vehicle cost.

Estimates of the vehicle on–cost when using alcohol as a fuel are most frequently obtained from American studies, since ethanol and methanol fuel programmes have been underway in that country for many years. The US EPA (1989) have presented cost estimates provided by the Ford Motor Company for FFVs able to use M85 as a fuel. For a vehicle produced in high volume (100,000+ per annum), the incremental cost over that for a petrol–fuelled vehicle is $200 to $400. The California Air Resources Board (CARB) estimate the added costs for methanol–fuelled light–duty vehicles to be in the range $200 to $370 in order to comply with increasingly stringent emissions legislation (from the California TLEV through to ULEV standards). Costs for ethanol–fuelled equivalent vehicles are expected to be roughly the same as for the methanol figures quoted above.

Limited production prototype FFVs that have been demonstrated in the US have cost significantly more than their petrol–fuelled counterparts. The US General Accounting Office (GAO, 1991) has summarised the incremental costs of procuring medium–sized alcohol–fuelled vehicles (65 methanol FFVs) in 1990. The average cost of the alternative fuel components was $2,250 and represents 30 percent of the typical price of a petrol–fuelled government–procured compact vehicle in 1990 ($7,730), or about 16 percent of the total price of the alternatively–fuelled vehicles themselves ($14,130 each).

Ford and General Motors have provided testimony to the US Congress stating that for dedicated alcohol vehicles produced in volumes of at least 100,000 per annum, the cost would be no more than that of a comparable petrol vehicle (EPA, 1989). The extra costs of more sophisticated fuel injection equipment (to help cold starting), ethanol and methanol–compatible components, larger fuel tank and aldehyde emission control components are expected to be offset by the use of a smaller, lighter engine (and compound weight saving effects), the requirement for a smaller engine cooling capacity and reduced evaporative emission controls.

4.8   Demonstration

The use of ethanol as a vehicle fuel in Brazil arose as a result of economic pressures in the early to mid-1970s. Brazil was spending over one third of its export earnings on petroleum imports and when, in 1975, the worldwide sugar market slumped, it seemed logical to use an indigenous product as a fuel supply. Thus began ProAlcool, the National Alcohol Program to support the sugar industry and reduce oil imports (Sathaye *et al*, 1988). The initial phase focused on increasing the ethanol percentage in gasohol to 20 percent nationwide and in 1979 the second phase involved a major shift to produce and supply dedicated ethanol vehicles. Not only do four million Brazilian cars now use ethanol as a fuel but an additional nine million run on a 20–22 percent blend of alcohol and petrol. Currently 10 companies manufacture FFVs capable of using mixtures of petrol and up to 85 percent ethanol (E85). The technology is based on experience with neat ethanol vehicles in Brazil and methanol FFVs in the USA.

Probably the world's largest ethanol–fuelled bus trials are being conducted over a three year period in Stockholm. SL (the Stockholm Transport Authority), Saab–Scania and the Swedish Ethanol Development Foundation are running the trials, involving 32 buses, fuelled with ethanol containing ICI's "Avocet" ignition enhancer (28g/kg fuel) and a corrosion inhibitor. The buses, fitted with Saab–Scania 184 kW modified DSI 11E diesel engines and oxidation catalytic converters are claimed to emit very low levels of HC, CO and particulates, with NO_x emissions able to satisfy US 1994 urban bus emissions standards (Storstockholms Lokaltrafik, 1990). SIP (1993) reports that emissions from second–generation Saab–Scania ethanol engines according to ECE R49 testing are expected to be 0.10 g/kWh CO, 0.09 g/kWh HC, 3.8 g/kWh NO_x and 0.05 g/kWh particulates. Originally planned to finish in 1993, the Stockholm trial has been extended.

Detroit Diesel Corporation has installed a 224 kW 6V–92TA heavy–duty modified diesel engine, running on ethanol, in a tractor unit for use by a Colorado Brewery (Professional Engineering, 1991). The ethanol is produced as a byproduct of the brewing process. Ethanol–fuelled bus demonstration projects are also occurring in several US and Canadian cities. Most automotive manufacturers that have been engaged in research and development with methanol as a fuel have also included ethanol in their activities. Many aspects of the R&D and possible production planning of dedicated, converted or flexible fuel vehicles have been with alcohols as a generic fuel – thereby including both methanol and ethanol.

4.9   Outlook

Ethanol is, in several ways, an attractive automotive fuel, likely to find applications in niche markets or geographical areas with a favourable feedstock supply (and other factors, such as local constraints on fossil fuel use), for example in Brazil. It is likely to provide some emissions benefits over petrol, though the benefits of neat ethanol (E100), or ethanol blended with small amounts of petrol, must be considered uncertain because of a lack of experience with vehicles equipped with modern emission control equipment. It is basically a safer fuel than petrol to distribute and use, it is in convenient liquid form and its volumetric energy content is higher than the other leading alternative fuel contenders, minimising range limitation problems.

The major barrier to its introduction and widespread use is fuel (and feedstock) supply and the cost of growing crops and producing the fuel. Prospects are therefore not favourable for substantial increases in ethanol use in transport applications, relying on the current ethanol production systems. Short-term improvements in the efficiency of present systems could enhance ethanol's cost and energy balance, but seem unlikely to provide a boost necessary for major production increases. For the long term – beyond 2000, ethanol's prospects may improve given the potential for relatively inexpensive production from wood and wastes, based on the enzymatic hydrolysis process currently being actively developed.

4.10 Summary

Ethanol is a liquid fuel with a high octane rating (RON=108) but a lower energy density than petrol or diesel (requiring a volume of 1.5, falling to perhaps 1.25 times that of petrol for equivalent vehicle range). Ethanol has a low cetane rating, thereby requiring spark or ignition assistance or a fuel additive in a diesel engine.

As a petrol substitute

↓ evap. emissions, ↓ HC reactivity (but less reduction than for methanol), CO variable but no significant difference.

NO_x emissions variable: ↑ with increased compression ratio (dedicated engine), can be much ↓ in FFVs. Using $\lambda=1$ with 3WC will produce NO_x emissions comparable with petrol.

much ↑ acetaldehyde, although much ↓ benzene, ↓ 1,3-butadiene.

As a diesel substitute

substantially ↓ particulates (80% lower).

↓ NO_x.

HC and CO emissions variable.

↓ PAH.

REFERENCES FOR CHAPTER 4

CRAGG C (1992). Cleaning up motor car pollution: New fuels and technology. Financial Times Management Report, FT Business Information Ltd., London.

DELUCHI M A (1991). Emissions of greenhouse gases from the use of transportation fuels and electricity. Technical Memorandum No ANL/ESD/TM-22. Argonne National Laboratory, Argonne, Illinois, United States of America.

DE OLIVEIRA A (1991). Reassessing the Brazilian alcohol programme. Energy Policy, Pages 47-55, January/February 1991. Butterworth-Heinemann Ltd.

ENVIRONMENTAL DATA SERVICES LIMITED (1992). Brussels and the bioethanol boondoggle. Pages 22-25, ENDS Report 212, September 1992.

HALL D O, F ROSILLO-CALLE and P DE GROOT (1992). Biomass energy: Lessons from case studies in developing countries. Energy Policy, Pages 62-73, January 1992. Butterworth-Heinemann Ltd.

HOLMAN C, M FERGUSSON and C MITCHELL (1991). Road transport and air pollution: Future prospects. Rees Jeffreys Discussion Paper 25, Transport Studies Unit, Oxford University.

JOSEPHSON J (1992). Biofuel development. Environ Sci Technol, Pages 660-663, Vol 26, No 4, 1992.

NIEMI S and E UBONG (1991). Exhaust emissions of an ethanol DI spark-assisted diesel engine equipped with a catalytic converter. SAE Technical Paper Series No 910850, Society of Automotive Engineers, Inc., Warrendale, Pennsylvania, United States of America.

PITSTICK M E (1993). Emissions from ethanol and LPG fueled vehicles. Preprint paper No 930527, Transportation Research Board 72nd Annual Meeting, 10-14 January 1993, Washington, DC, United States of America.

PROFESSIONAL ENGINEERING (1991). Transport fuels: Making motors change their tipple. Pages 30-31, April 1991.

SATHAYE J, B ATKINSON and S MEYERS (1988). Alternative fuels assessment: The international experience. International Energy Studies Group, Lawrence Berkeley Laboratory, University of California, Berkeley, California, United States of America.

SIP (1993). Transport & Vehicles: Successful tests with Ethanol–driven buses completed in downtown Stockholm. Press Release, May 1993. SIP – The Swedish International Press Bureau, Stockholm, Sweden.

STORSTOCKHOLMS LOKALTRAFIK AB (1990). Ethanol: An alternative? Information leaflet. SL, Stockholm, Sweden.

SZWARC A and G M BRANCO (1985). Automotive use of alcohol in Brazil and air pollution related aspects. SAE Technical Paper Series No 850390, Society of Automotive Engineers, Inc., Warrendale, Pennsylvania, United States of America.

THE ENGINEER (1992). Environment: Bug turns paper mill and crop wastes into petrol additive. Page 38, 7 May 1992.

US CONGRESS (1990). Replacing gasoline: Alternative fuels for light–duty vehicles. Office of Technology Assessment, US Government Printing Office, Washington, DC, United States of America.

US DEPARTMENT OF ENERGY (1991). Technology Factsheet No 10.0: Biofuels research. US DOE, Office of Transportation Technologies, Washington, DC, United States of America.

US ENVIRONMENTAL PROTECTION AGENCY (1989). Analysis of the economic and environmental effects of ethanol as a motor fuel. Special Report, Office of Mobile Sources, EPA, Washington, DC, United States of America.

US GENERAL ACCOUNTING OFFICE (1991). Alternative fuels: Increasing Federal procurement of alternative–fuelled vehicles. US GAO, Report to Congressional Requesters, GAO/RCED–91–169, Washington, DC, United States of America.

WATERS M H L (1992). Road vehicle fuel economy. TRRL state–of–the–art review number 3. HMSO: London.

WRIGHT L L (Undated). Biofuels feedstock production overview. Oak Ridge National Laboratory, Oak Ridge, United States of America.

ZELENKA P, P KAPUS and L A MIKULIC (1991). Development and optimisation of methanol fuelled compression ignition engines for passenger cars and light duty trucks. SAE Technical Paper Series No 910851, Society of Automotive Engineers, Inc., Warrendale, Pennsylvania, United States of America.

ZELENKA P (1991). Alcohol fuels for internal combustion engines. Proceedings of the conference "Engine and Environment – Which Fuel for the Future?", 23–24 July 1991, Grazer Congress, Graz, Austria.

5. BIODIESEL AND VEGETABLE OILS

5.1 Introduction

Vegetable oils, such as rapeseed, linseed, cottonseed, soybean, sunflower, castor, peanut, coconut, palm and others, are candidates for alternative fuels in diesel engines. The fuel crises of the 1970s and early 1980s focused attention on the desirability to develop alternative (renewable) fuels and decrease the dependency on fossil fuel reserves. However, as early as 1900 Rudolph Diesel's first engine was able to burn vegetable oils as well as animal fats and fossil fuels. During the Second World War many vehicles, primarily in southern France, used vegetable oil fuel substitutes (Andrzejewski, 1991).

Today the environmental benefits from the use of bio–fuels are promoted as justification for the growing of the feedstock crops. While the theoretical ecological advantage of these fuels, namely their CO_2–neutrality (that is, as much CO_2 being consumed during the growth of the plants as being generated through their combustion) are not disputed, the overall CO_2 balance may be unfavourable with regard to the greenhouse effect, according to some studies, while others claim a significant advantage over fossil fuels.

Most vegetable oils are able to be substituted directly for diesel fuel, but may create a variety of practical problems largely resulting from incomplete combustion but also causing injector nozzle coking and even failure, excessive engine deposits, lubrication oil dilution, piston ring sticking, scuffing of the cylinder liners and even lubricant failure due to polymerisation of the vegetable oil (Knothe *et al*, 1992). Other operational factors such as poor cold starting, unreliable ignition and misfire and reduced thermal efficiency (with certain oils) have added to the general avoidance of unmodified vegetable oils as a long–term diesel fuel replacement, especially in DI diesel engines and small–capacity IDI diesel engines where the detrimental effects have been greatest.

A more satisfactory fuel is obtained by chemically changing the vegetable oil into a form that more closely resembles the diesel fuel it is designed to substitute. A well known example is the transesterification (transformation of the oil into fatty acid ethyl esters) of vegetable oils, with rapeseed oil methyl ester (RME) being perhaps the most common form of "bio–diesel".

5.2 Fuel characteristics

Most published information on the testing of vegetable oils is for RME. Discussion of the various aspects of the use of vegetable oils as a vehicle fuel in the remainder of this chapter are based on those aspects of RME, except where specifically expressed otherwise.

Pure rapeseed oil and RME differ from diesel fuel in their chemical structure, but RME more closely resembles diesel in its physical properties and is able to be used as a direct fuel substitute with no vehicle or engine modification usually required. The most striking difference between diesel and rapeseed oil and RME is that the latter comprise about 11 percent chemically bound oxygen. The average molecular mass is about the same for diesel and RME but much higher for pure rapeseed oil. While there is some loss in energy output with RME, generally in practical use no difference is observed (Pächter and Hohl, 1991). RME possesses better ignition quality than diesel fuel with a cetane rating of 54–58 compared with 50–54 for diesel fuel and 40–44 for pure rapeseed oil (Richter *et al*, 1991). In addition, RME is mixable with conventional diesel fuel in any ratio.

Experience of material compatibility with RME has so far been good. Rubber and polymer–based components are thought to be the critical items, but proper material specification appears to offer good resistance to the fuel. Some paints, however, are thought to be susceptible to corrosive attack.

One significant advantage of RME and other vegetable oils is that they are renewable. Fossil fuel reserves are finite and pressures to develop renewable energy forms are growing. A number of countries, especially in the industrialised western world, are faced with overproduction in agriculture. The use of surplus agricultural crops to produce a diesel substitute may or may not help reduce the threat of an enhanced greenhouse effect, but would make some contribution to extending conventional fossil fuel supplies. The International Energy Agency (1994) estimates that RME has about 35 percent lower life–cycle energy use than diesel fuel and total life cycle CO_2 emissions of 50 percent. Other studies claim figures below, and in excess of this quantity.

The calorific value of RME is about 12 percent lower than that of diesel fuel but this is partly compensated by the seven percent higher density of RME. The total loss in energy output (on a volume basis) is about five percent (Fischler, 1991). Other testing of tractor engines (Macchi, 1991) has showed that a mean power reduction of around 3.5 percent is obtained – only the worst cases are at the five percent mark.

The ignition delay of the majority of vegetable oils is longer than that for diesel fuel and so some change to the fuel injection timing would normally be required when using vegetable oils as a diesel substitute (Andrzejewski and Sapinski, 1991). The ignition delay of RME is much shorter than pure vegetable oils, and very close to that of diesel fuel.

The viscosity of RME is slightly higher than diesel fuel (kinematic viscosity at 20°C, v is 6–8 cSt, compared with 3.7 cSt for diesel), but significantly lower than that of rapeseed oil (68–75 cSt) and other pure vegetable oils (Richter *et al*, 1991). Vegetable oils are also more corrosive than diesel fuel and their use has lead to injector nozzle coking and failure, excessive engine deposits, lubrication oil dilution, piston ring sticking, scuffing of the cylinder liners and even lubricant failure due to polymerisation of the vegetable oil (Knothe *et al*, 1992).

The effect of dilution of the lubricating oil with the use of RME is cited as probably the most important concern with regard to engine durability. Direct contact between oil and RME occurs in the fuel injection pump at the plunger and there is also contact during the fuel injection/combustion process in the cylinder. Experimental investigation both in field tests and test bed trials have shown lubrication oil dilution of 4–5 percent on average, with worst cases around eight percent. Kaiser (1991) reports that adjustment of the fuel injection pump can be made to help reduce lubrication oil dilution.

It is believed that oil dilution increases with engine age, mostly because more fuel is able to enter the lubrication system. With diesel that enters the lubrication oil, a major part of the fuel evaporates on account of the high content of components with low boiling points (diesel boils within a wide temperature range between 180 and 360°C), but this is not possible to such an extent with RME due to the ester molecules being quite similar and all boiling at about the same temperature of 360°C (Pächter and Hohl, 1991).

Carbon deposits, especially on inlet valves, have traditionally been a problem with neat vegetable oils, but have also been found to be a problem with RME. Macchi (1991) reports that during 1,000 hours of engine operation on RME, 2mm thick deposits accumulated on the engine inlet valves. One specific problem appears to be with back–flow of the exhaust gases through the inlet valves, prior to incoming air entering the cylinder. Further investigation of this phenomenon is underway, and specific RME fuel additives may be developed to help prevent this occurring.

Low–temperature behaviour of RME is less favourable than that of diesel fuel and fuel additives may need to be developed or fuel heating employed in certain operating environments. The typical cold filter plugging point (CFPP) of RME is −9°C, compared with −22°C for conventional diesel fuel.

5.2.1 Safety implications

RME degrades rapidly if spilled on soil – 98 percent is decomposed within three weeks. This would appear to offer RME the prospect of being considered as a candidate fuel for application in ecologically sensitive areas and for vehicle operation activities potentially hazardous to the local environment (Fischler, 1991).

5.3 Feedstocks

The individual feedstocks have already been mentioned, and include rapeseed, linseed, cottonseed, soybean, sunflower, peanut, coconut and palm. For a number of reasons, rape has proved to be the most favourable in central Europe. Rape has several advantages as a candidate feedstock for an alternative fuel. The crop yield is relatively high – in the region of 1,300 litres of fuel/hectare and is superior to all other oil plants (Pächter and Hohl, 1991). According to experts, annual rape yields are expected to increase at more than three percent for the foreseeable future. Rape has a high resistance to attack by other plants and animals and generally requires less treatment with herbicides and pesticides.

The production of RME is a two–step process. The oil is first extracted by pressing the crop – the liquid oil fraction goes on to be converted and the remaining oil cake (with a 6–8 percent oil content) is used as a high–quality protein animal feed. The extracted oil, after filtering, is transesterified, where the trivalent alcohol glycerin is replaced by three molecules of monovalent alcohol. This process, called alcoholysis, is based on the so-called Bradshaw method, although more sophisticated processes are currently being developed (Mittelbach, 1991). Usually the alcohol added is methanol, which gives rapeseed oil methyl ester and releases glycerin, under the action of an alkaline catalyst.

The glycerin byproduct can be processed to purify it and is able to be used by the pharmaceutical and cosmetics industry, although significant further increases in the amount of glycerine produced would quickly lead to a surplus. The separated methanol is conditioned and returned to the process. 1.05 tonnes of rapeseed oil plus 110 kg of methanol will yield 1 tonne of RME plus 100 kg of glycerine, plus fatty acids, according to the French Environment Agency (ADEME, 1992).

5.4   Infrastructure

Supporters of bio–fuels would never claim that conventional fossil fuels can be entirely replaced with RME or other fuels – the land area required would be colossal and displace needed food crops. The role for bio–fuels, it appears, may be to fill niche markets and provide a partial substitute for conventional fuels. Such an example would be to use locally–grown rapeseed to produce RME to fuel farm equipment or to use local feedstocks to fuel vehicles in remote areas – especially developing countries.

The distribution, storage and refuelling infrastructure for vegetable oils should not be very different to that for diesel fuel. Water ingress should be prevented and general degradation avoided. A greater volume of RME is required to maintain vehicle performance and range relative to diesel, demonstrated at between 1.3 and 18 percent depending upon vehicle application and speed, with most testing revealing a volumetric increase of about five percent required. Experience to date with extended periods of storage of vegetable oils and transesterified oils has revealed no concerns (Kaiser, 1991), although cold temperatures may have an adverse effect (restricted flow due to high viscosity) unless necessary additives or fuel heating are used.

5.5   Vehicle modifications

The advantage quoted for the use of RME as a diesel fuel substitute is that it can be substituted directly without the need for vehicle or engine modifications. Some material re–specification, especially for rubber and polymer components, may be advantageous in the long term. Many pure vegetable oils can be used with little or no modification, but usually with serious consequences due to their high viscosity and acidity.

5.6 Emissions performance

5.6.1 Vehicle exhaust emissions

5.6.1.1 Methyl esters

Most emissions testing of vehicles using RME report that the major advantage is the very low sulphur dioxide (SO_2) emissions due to the fuel containing very little sulphur. Regulated exhaust emissions results have shown considerable variation, depending on the engine condition, vehicle application and test cycles employed. These factors may cause greater differences in exhaust emission rates than may be attributed to the use of bio-diesel itself.

Pächter and Hohl (1991) have reported on fleet trials in the Austrian army, but their test programme used the ECE R49 13–mode engine tests (rather than vehicle road tests) to determine the relative emissions performance. Their results showed HC emissions from the use of RME to be 50 percent lower than from diesel fuel, with NO_x slightly higher and CO slightly lower. A notable point is the variation of emission rate with engine mean effective pressure (mep) in which HC emissions between RME and diesel fuel vary considerably more than do the NO_x and CO emissions. This implies that the vehicle duty cycle will play a major role in determining certain exhaust emission characteristics.

Novamont (1992) present exhaust emissions test results from the use of RME and SME (soybean methyl ester) in a heavy–duty diesel engine (DDC 6V–92TAC two–stroke bus engine), tested according to the US EPA transient cycle with low–sulphur (0.05%) diesel as the reference. The results are shown in Table 9.

Exhaust emissions g/bhp–hr	Diesel (0.05% sulphur)	RME (% diesel emissions)	SME (% diesel emissions)	SME with methanol catalyst	SME with diesel catalyst
NO_x	4.840	5.614 (116)	5.787 (120)	5.677 (117)	5.728 (118)
HC	0.437	0.093 (21)	0.116 (27)	–	0.068 (16)
CO	1.507	0.811 (54)	0.873 (58)	0.141 (9)	0.874 (58)
Particulates	0.227	0.164 (72)	0.152 (67)	0.054 (24)	0.062 (27)
CO_2	758.1	775.5 (102)	791.3 (104)	805.9 (106)	794.0 (105)

Table 9.   Exhaust emission results from US EPA heavy–duty transient testing of rapeseed and soybean methyl esters

The results include untreated RME and SME emissions and those from the use of soybean methyl ester with two types of oxidation catalyst. The results are very consistent for each fuel compared with the diesel reference. NO_x emissions are all higher (by about 20 percent) and HC, CO and particulate emissions are lower than with diesel (raw HC emissions are about one–fifth to one–quarter, CO emissions are about half and particulate emissions are about two–thirds). The oxidation catalyst appeared not to make a significant difference in HC levels with the use of SME, whereas the CO emissions fell substantially with a methanol catalyst when using SME fuel. The particulate emissions fell to around one–quarter of those from diesel when an oxidation catalyst was used with SME fuel. However, it should be noted that a fairer comparison would have been obtained if the diesel was also tested in conjunction with an oxidation catalyst.

Tests conducted during 1985 by the Institut du Pétrole (IFP) on an MWM heavy–duty DI diesel engine according to a 13–mode cycle (believed to be the now obsolete US 13–mode steady state procedure – similar to ECE R49) and tests in 1988 on a Peugeot XUD9 passenger car diesel engine according to the ECE R15 cycle, using diesel fuel and RME are reported by the French Environment Agency (ADEME, 1992). Tests by UTAC on a Renault diesel bus engine (RVI PR 100.2) according a 13–mode cycle and another test cycle – AQA, using RME and diesel fuel are reported and also presented in Table 10.

Exhaust emissions g/kWh, g/km or g/test	Renault bus engine RVI PR 100.2				MWM engine: 13–mode cycle		Peugeot XUD-9 engine: ECE R15 cycle (g/test)	
	13–mode cycle		AQA (g/km)					
	Diesel	RME (%dies)	Diesel	RME (%dies)	Diesel	RME (%dies)	Diesel	RME (%dies)
NO_x	17.90	19.60 (109)	21.82	20.38 (93)	4.60	5.30 (115)	4.19	5.02 (120)
HC	0.62	0.48 (77)	2.15	1.55 (72)	1.60	1.40 (88)	1.28	0.85 (66)
CO	2.53	1.88 (74)	5.87	5.26 (90)	8.00	2.70 (34)	4.64	5.38 (116)
Particulates	0.801	0.496 (62)	1.19	1.13 (95)	–	–	0.67	0.39 (58)

Table 10. Exhaust emission results from IFP and UTAC testing of heavy and light–duty diesel engines fuelled with RME

The results in Table 10 show the same trend as those in Table 9. NO_x emissions are up to one–fifth higher with RME than diesel and HC, CO and particulate emissions are lower by varying amounts. The passenger car diesel engine (Peugeot XUD-9), however, gave a higher CO emission value with the RME, which contrasts with the heavy–duty engines.

The Swiss Federal Laboratories for Materials Testing and Research (EMPA) have undertaken testing of a 12–litre, 184 kW, 6–cylinder Mercedes–Benz diesel engine using diesel, low–sulphur diesel (0.06%), sulphur–free diesel (<0.01%) and Austrian and Italian–produced RME fuels (Walter, 1992) according to the ECE R49 test procedure. The test results reveal specific fuel consumption (g/kWh) of the RME to be about 12 percent higher than normal diesel fuel, and engine power lower by between 1.8 and 7.5 percent.

Relative to normal diesel, the RME fuel produced exhaust emissions of CO 33 percent lower, NO_x 14 percent higher, HC about five percent higher and particulate emissions about five percent higher. Notably, testing of the engine with the various fuels using an oxidation catalyst, particulate filter and catalyst/filter combination, indicated that use of the RME fuel generally gave the lowest exhaust emission levels. With an oxidation catalyst/particulate filter combination, the CO emissions were reduced by 80 percent, NO_x was unchanged, HC emissions were 75 percent lower and particulates some 90 percent lower than the engine using normal diesel fuel with no catalyst or filter.

Testing by AVL in Austria using RME in prototype 2.3–litre turbocharged DI diesel engines (with EGR) under transient chassis dynamometer operation (US–FTP 75) is summarised in Figure 12 (from Tritthart and Zelenka, 1990). HC, CO and particulate emissions were lower (by 20%, 13% and 16% respectively) than those from diesel, whereas NO_x increased by about 20 percent. The organic insoluble portion of the particulates was significantly decreased, mostly due to the oxygen content of the fuel (about 10% by weight) permitting better combustion and soot oxidation. Better combustion and a higher temperature is the main reason for higher NO_x emissions.

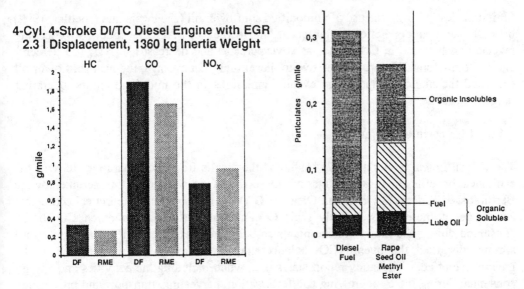

Figure 12. Exhaust emissions results from a RME–fuelled 2.3–litre diesel engine

Volumetric fuel consumption of the RME was 9 percent higher than the diesel fuel, although on an *energy basis,* the consumption was identical. Observed Polycyclic Aromatic Hydrocarbon (PAH) emissions were 10 percent lower with RME, but were not considered statistically significant due to large standard deviations in PAH analysis. Total aldehyde emissions were 25 percent higher with RME (57 mg/mile compared with 43 mg/mile).

5.6.1.2   Unmodified vegetable oils

Richter *et al* (1991) report on test results from the operation of a range of diesel engines (from car to tractor to heavy goods) on unmodified rapeseed oil using both ECE R49 and ECE R15 test cycles, depending upon the engine application. With the heavy duty engines tested according to ECE R49, HC emissions were about 1½ times those from diesel fuel, NO_x emissions were up to 20 percent lower, CO was 1½ to 2 times and particulate emissions up to twice those from diesel fuel with the DI engines while the IDI engines produced lower levels (30–50 percent reduction) of particulate emissions.

The passenger car diesel engine tested with rapeseed oil demonstrated HC emissions 2½ times those from diesel, NO_x emissions 15 percent lower, CO three times and particulate emissions almost three times those from diesel fuel. Aldehyde emissions were between three and four times those from diesel. Hemmerlein *et al* (1991) reports an update of some of the results presented by Richter *et al* (1991) above, with the most notable item being that PAH emissions were reduced with rapeseed oil with large–capacity IDI engines, but rose with DI engines and small IDI diesel engines.

Emissions testing from the use of unmodified sunflower oil (Ziejewski and Goettler, 1992) in a DI one–cylinder Petter engine according to the US SAE 13–mode test procedure revealed an increase in CO emissions (five percent) and decreases in NO_x (45 percent) and HC emissions (35 percent) relative to diesel fuel. On average, the neat sunflower oil exhibited the highest reduction in exhaust emissions in the mid–load engine operating range.

5.6.2 Life cycle emissions

The British Embassy Bonn (1992) has noted the results from a comparative study of the ecological benefits and disadvantages of rapeseed oil versus diesel fuel, undertaken by the Berlin–based Federal Environment Office (UBA). Even though the theoretical ecological advantage of these fuels, namely their CO_2–neutrality (that is, as much CO_2 being consumed during the growth of the plants as being generated through their combustion) are not disputed, the overall CO_2 balance may be unfavourable with regard to the greenhouse effect. The study report states that, when including the activities and energy consumed during the crop growing and fertilisation, harvesting, transport and processing, the use of rapeseed oil would increase global warming. This study also concludes that while SO_2 and NO_x emissions are very low compared with diesel, CO and HC emissions are significantly higher and the exhaust odour is unpleasant.

The UBA study of the RME fuel life cycle CO_2 emissions, quoted above, has been criticised by French researchers who claim that RME produces much less CO_2 than the fossil fuel it replaces (New Scientist, 1992b). Scientists at France's Environment and Energy Agency, (ADEME, 1992), found that the growing, processing and burning of RME produces one–fifth as much greenhouse gas emissions as fossil fuels (or rather the total energy yield of RME is five units for one unit of expended fossil fuel energy). The ADEME estimate of a 5:1 energy ratio compares with a range of 1.3:1 to 3.8:1 by Culshaw and Butler (1992), depending on the use of by–products.

The French claim that differences in calculation methods and the use of different global warming potentials (GWPs) for various greenhouse gases caused the variation in conclusions, combined with the claim that the German study overestimated the amount of nitrogen oxides emitted from soil as a result of the use of nitrogen fertilisers. The sharply contrasting conclusions from these two studies seem unlikely to be caused simply by differences in calculation methods. It should be noted that energy credit appears to be given for by–products (such as oilcake) obtained from RME production, but it may be unwise to assume that a market would exist for all the by–products, especially as RME production capacity increases significantly over the next few years as predicted. The total energy yield of the RME (*without* credit for by–products) is double that of the fossil fuel energy expended, or put another way, RME produces about 50 percent as many greenhouse gas emissions as fossil fuel, according to ADEME.

The International Energy Agency (1994) estimates that using fossil fuels for the processing energy and current technical practice, RME has about 35 percent lower life–cycle energy use than diesel fuel. CO_2 emissions are believed to be reduced by more than this because of the (increasing) use of natural gas and non–fossil fired electricity sources to provide the process energy – the relative life–cycle CO_2 emissions from RME are 50 percent.

5.7   Costs

5.7.1 Fuel production and distribution

Earlier estimates of the costs of producing bio–diesel in Western Europe have ranged from four and eight times that to produce diesel fuel, depending on the level of agricultural subsidies (Seiffert and Walzer, 1991). Recent estimates are still variable, but the increase over the production cost of diesel appears much less that previously quoted. Some Austrian producers claim they can make it for the same price as diesel (New Scientist, 1992a), but the fuel is generally only competitive with a tax subsidy at the filling pump.

The French Environment Agency (ADEME, 1992) show that the cost of producing vegetable oil fuels range from 3 to 2½ times that of diesel, in the short to medium term. The French "TIPP" (domestic petroleum products consumption tax) enables the pump purchase price of vegetable oil fuels to fall to around 1.2 to 1.0 times that of diesel, again in the short to medium term.

A review of the potential of biodiesel as a transport fuel for the UK Department of Trade and Industry (Culshaw and Butler, 1992) concluded that at the existing market price for rapeseed, the cost of untaxed biodiesel, including credits for the glycerine and livestock meal by–products, would be approximately 26p/litre. To make biodiesel competitive with untaxed post–refinery diesel fuel (about 11p/litre), a subsidy of about 15p/litre would be required or the cost of diesel would have to increase by 240 percent.

The pump price of the RME purchased for the Reading Buses trial was comparable with that of diesel, which is taxed, at the time of purchase (Autumn 1992). Because the bus company receives a rebate for the duty paid on diesel, the RME was reported as costing between two and three times more per gallon of diesel fuel (Independent, 1992). New EC regulations from 1 January 1993 (Independent, 1992, Diesel Car, 1992) will mean that duty will be imposed on diesel substitutes at the prevailing diesel rate. Future EC excise duties on biofuels (proposed in the Scrivener Directive at 10 percent of the fossil fuel replaced) appear uncertain, with a significant international lobby against reducing duty.

5.7.2   Vehicle modification

No vehicle modification is necessary with the use of RME as a diesel fuel substitute and consequently no cost for any modification is incurred when a vehicle is operated on this fuel. Technically a small increase in the capacity of the fuel storage system would be required to maintain comparable vehicle operating range with RME. Some sources indicate that the use of RME in older engines may necessitate some engine and fuel component recalibration.

The use of pure vegetable oils would require some modification – if only an adjustment to the fuel injection timing to allow for the longer ignition delay of the oil compared with diesel fuel. To help maintain engine durability, material specifications may need revision to resist the increased acidic and corrosive effects of vegetable oils over those from diesel, but these should only impose a relatively small increased engine/vehicle cost, especially as pure vegetable oils are more suitable for use in large capacity diesel engines.

5.8   Demonstration

Extensive use of RME as a diesel fuel substitute has taken place in Austria and Italy over the past few years. The first RME production plant in Austria was opened in August 1987. Fendt Tractors have undertaken fleet trials of RME with over 100 tractors in use and S+L+H (Same, Lamborghini, Hurlimann – a group of companies making tractors and agricultural diesel engines) have also held field trials involving over 100 units. Extensive fleet trials of RME as a diesel fuel alternative have been undertaken by the Austrian Army.

Much interest in the use of bio–fuels has prompted French organisations to undertake fleet trials and engine tests using vegetable oils and methyl esters for several years. Many trials have involved the use of RME/diesel blends in varying quantities. The Renault engine test results reported in section **5.6.1.1** were from a bus operated by the transport company of the City of Tours, undergoing a "green fuel" trial.

In the UK, much attention (Independent, 1992, New Scientist, 1992a) has been paid to Reading Buses' use of RME as a diesel fuel substitute for a three–month trial period, commencing in November 1992. Initial plans included the monitoring of a double–decker and single–decker bus and a coach, to see how they perform on the fuel. Reading Buses (part of Reading Transport) is believed to be the first British bus company to use RME as a diesel alternative. The initial 22,000 litres of fuel was obtained from the Italian supplier Novamont, part of the Ferruzzi–Montedison Group, who also supply 17 Italian bus companies, the Zurich municipal bus company, the South Dakota bus company and the Berlin taxi association with RME.

Mercedes–Benz has tested a model 190 passenger car able to be fuelled with a range of vegetable oils, although designed primarily for RME (Autocar & Motor, 1992). The German agricultural company Hoechst claims that in many cases the vehicle performance is better than with diesel fuel. However, the vehicle costs twice the price of a conventional diesel 190, mostly because of the specially–designed turbocharged, direct injection multi–fuel engine. Thermo King is testing rapeseed oil as a diesel fuel alternative for its refrigeration equipment mounted on heavy goods vehicles (Commercial Motor, 1992).

5.9   Outlook

There are several reasons to believe that vegetable oils will play an increasingly greater role as a substitute for conventional diesel fuel in the future. Fossil fuel reserves are finite and public and political pressure to develop renewable energy forms are growing. A number of countries, especially in the industrialised western world, are faced with overproduction in agriculture. The use of surplus agricultural crops to produce a diesel substitute – "bio–diesel" may or may not help reduce the threat of an enhanced greenhouse effect, but would make some contribution to extending conventional fossil fuel supplies.

Reforms made in May 1992 to the Common Agricultural Policy stipulated that arable farmers must take 15 percent of their land out of food production. The Scottish

Agricultural College in Edinburgh has been commissioned by Scottish Enterprise, to decide the best alternative uses for rapeseed oil, including its possible niche market for a diesel fuel substitute (Daily Telegraph, 1992). Scotland has been chosen for the research because of its favourable growing conditions compared with the rest of the UK. It has been estimated that half of the tractors in Scotland could be run on the 450,000 tonnes of rapeseed harvested each year from the 150,000 acres of rape grown there (ibid). In the UK as a whole, if all the estimated 630,000 hectares of set–aside land (in 1992) were used for RME production, 6.4 percent of the total diesel fuel market would be replaced (Culshaw and Butler, 1992).

Austria has several small RME production plants processing rape crop from 500 to 2,000 hectares each annually, and two large–scale commercial plants are able to jointly process the crop from 25,000 hectares annually. During 1992 Austria hopes to be able to market 23,000 tonnes of bio–diesel (RME) in that country (Fischler, 1991). Companies in France, Germany, Italy and Spain are also building RME production plants and some commentators have estimated that production of RME could be 600,000 tonnes per annum in Europe within five years (New Scientist, 1992a).

RME is still a relatively new fuel and further research and development needs to be conducted before it could be used on a larger scale, or for general public use. The full environmental effects of the use of bio–diesel needs closer evaluation, both in terms of the energy balance and vehicular exhaust emissions performance in different engine/vehicle applications and duty cycles. An EC proposal for a biodiesel specification has been made and requires agreement and approval, particularly for specific fuel additives to provide the necessary level of engine and fuel component protection currently offered by diesel fuel. In Austria, substantial work on a standard for RME as a substitute for diesel fuel has taken place, and a fuel quality requirement standard (ÖNORM, or Austrian Standard C1190) was published in February 1991 (Mittelbach, 1991).

5.10   Summary

Advantages of RME

Comparable performance with diesel fuel, with better ignition quality/cetane rating. Blending with diesel is possible. Fuel degrades rapidly if spilled.

Disadvantages of RME

Lower calorific value than diesel fuel, partly offset by higher density $\Rightarrow$ about 5% power loss. Higher viscosity than diesel, and higher cold filter plugging point. Potential durability problems with fuel dilution in engine oil and carbon deposits on inlet valves. Some rubber/polymer component incompatibility.

Emissions performance of RME

Light duty vehicles: ↑ NO_x (about 20%), ↓ HC (20–30%), CO variable, ↓ particulates (20–40%).

Heavy–duty engines: ↑ NO_x (up to 20%), ↓ HC (20–75%), ↓ CO (10–50%), ↓ particulates (5–40%).

REFERENCES FOR CHAPTER 5

AGENCE DE L'ENVIRONNEMENT ET DE LA MAITRISE DE L'ÉNERGIE (1992). The use of vegetable oils or their by-products in diesel engines. Note No 1415, February 1992, Paris.

ANDRZEJEWSKI J and A SAPINSKI (1991). Some particularities of the vegetable oils combustion in diesel engines. Proceedings of the conference "Engine and Environment – Which Fuel for the Future?", 23–24 July 1991, Grazer Congress, Graz, Austria.

AUTOCAR & MOTOR (1992). News: 100 mph claim for veggie Mercedes. Page 8, 18 March 1992.

BRITISH EMBASSY BONN (1992). Rape seed versus diesel fuel – Federal Environment Office presents comparative study. Science & Technology Information Note number 194/92, 16 October 1992, Bonn, Germany.

COMMERCIAL MOTOR (1992). Vehicles News: Rape oil fuel in reefer test. Page 9, 26 November–2 December 1992.

CULSHAW F and C BUTLER (1992). A Review of the Potential of Biodiesel as a Transport Fuel. ETSU–R–71. Energy Technology Support Unit, Harwell.

DAILY TELEGRAPH (1992). Oilseed rape may fuel buses of the future. The Daily Telegraph, 25 July 1992.

DIESEL CAR (1992). News: Brussels boost for biodiesel. Pages 6–7, June 1992.

FISCHLER F (1991). Austrian farmers as producers of raw materials for biogenic fuels. Proceedings of the conference "Engine and Environment – Which Fuel for the Future?", 23–24 July 1991, Grazer Congress, Graz, Austria.

HEMMERLEIN N, V KORTE, H RICHTER and G SCHRÖDER (1991). Performance, exhaust emissions and durability of modern diesel engines running on rapeseed oil. SAE Technical Paper Series No 910848, Society of Automotive Engineers, Inc., Warrendale, Pennsylvania, United States of America.

IEA/OECD (1994). Biofuels. Un-numbered paper, January 1994. International Energy Agency/Organisation for Economic Co-operation and Development. Paris, France.

INDEPENDENT (1992). Buses to use rape seed fuel for trial period. The Independent, 3 November 1992.

KAISER W (1991). The use of alternative fuels in tractor engines – state of development and prospects from a manufacturer's viewpoint. Proceedings of the conference "Engine and Environment – Which Fuel for the Future?", 23–24 July 1991, Grazer Congress, Graz, Austria.

KNOTHE G, M O BAGBY, T W RYAN III, T J CALLAHAN and H G WHEELER (1992). Vegetable oils as alternative diesel fuels: Degradation of pure triglycerides during the precombustion phase in a reactor simulating a diesel engine. SAE Technical Paper Series No 920194, Society of Automotive Engineers, Inc., Warrendale, Pennsylvania, United States of America.

MACCHI S (1991). Overview on biodiesel utilisation for S+L+H tractor engines. Proceedings of the conference "Engine and Environment – Which Fuel for the Future?", 23–24 July 1991, Grazer Congress, Graz, Austria.

MITTELBACH M (1991). Methods for the transesterification of vegetable oils. Proceedings of the conference "Engine and Environment – Which Fuel for the Future?", 23–24 July 1991, Grazer Congress, Graz, Austria.

NEW SCIENTIST (1992a). Technology: British buses to run on flower power. Page 18, 3 October 1992.

NEW SCIENTIST (1992b). Technology: ...while row looms over biofuels. Page 21, 24 October 1992.

NOVAMONT (1992). The Novamont Company: Preliminary fuel evaluations by Detroit Diesel Corporation of rapeseed and soybean methyl esters. Information supplied by Ferruzzi Trading (UK) Ltd, on behalf of Novamont, to the Chief Mechanical Engineer's Office, Department of Transport.

PÄCHTER H and G HOHL (1991). Rapeseed oil methyl ester (RME) as an alternative diesel fuel: Fleet trials in the Austrian army. Proceedings of the conference "Engine and Environment – Which Fuel for the Future?", 23–24 July 1991, Grazer Congress, Graz, Austria.

RICHTER H, N HEMMERLEIN and V KORTE (1991). Use of rapeseed oil as an alternative fuel for diesel engines. Proceedings of the conference "Engine and Environment – Which Fuel for the Future?", 23–24 July 1991, Grazer Congress, Graz, Austria.

SEIFFERT U and P WALZER (1991). Automobile technology of the future. Society of Automotive Engineers, Inc., Warrendale, Pennsylvania, United States of America.

TRITTHART P and P ZELENKA (1990). Vegetable Oils and Alcohols – Additive Fuels for Diesel Engines. Proceedings from the XXIII FISITA Congress "The Promise of New Technology in the Automotive Industry", 7–11 May 1990, Turin, Italy.

WALTER T (1992). Untersuchung des Emissionsverhaltens eines Nutzfahrzeugmotors bei Betrieb mit Rapsölmethylester. EMPA No 133 439, Swiss Federal Laboratories for Materials Testing and Research, Dübendorf, Switzerland.

ZIEJEWSKI M and H J GOETTLER (1992). Comparative analysis of the exhaust emissions for vegetable oil based alternative fuels. SAE Technical Paper Series No 920195, Society of Automotive Engineers, Inc., Warrendale, Pennsylvania, United States of America.

6. LIQUEFIED PETROLEUM GAS

6.1 Introduction

Liquefied petroleum gas (LPG) has been in commercial motor fuel use worldwide for over 60 years. LPG can consist of propane or butane or mixtures of both, and unsaturated propylene and butylenes may be present in refinery LPGs (as opposed to liquefied gases from crude oil wells) and trace quantities of ethane or pentanes may also be present. Predominantly for supply reasons, automotive LPG sold in the USA is mostly propane, whereas in Europe it is usually a mixture of propane and butane, the proportions varying from country to country depending on local availability and ambient conditions. The propane content may be as low as 20 percent (such as in Italy) or as high as 90 percent in the United Kingdom or Germany. In the Netherlands, for example, the propane content would normally fall in the range 35–60 percent (Shell, 1976).

There are currently about four million LPG vehicles operating worldwide with about 700,000 in the Netherlands, over 500,000 in Italy, about 400,000 in the USA and over 100,000 in Canada (Pitstick, 1993). While most LPG vehicles use converted petrol engines, some original equipment manufacturer (OEM) light–duty vehicles were provided by Ford and Chrysler to the Canadian market in the early 1980s when LPG prices were low compared with petrol. Recently, Ford US announced the availability of medium–duty trucks specifically designed for LPG operation and Chrysler of Canada has announced that it intends to develop and produce LPG–fuelled light–duty trucks or vans by 1995 (ibid).

6.2 Fuel characteristics

LPG has a high octane rating of 112 RON (100% propane), enabling a higher compression ratio to be employed in a dedicated engine design and therefore improving the thermal efficiency. Relative fuel consumption of LPG, on a thermal basis, is about 90 percent of that of petrol (Shell, 1976). Because LPG has a simple chemical composition more complete combustion occurs, leading to lower CO and HC emissions.

Once the engine is running, full power is available immediately (for propane or propane/butane mixtures) because the LPG is already in a gaseous form, whereas for effective vaporisation of petrol, the inlet manifold has to be hot. The better fuel distribution between cylinders with a gaseous fuel usually results in smoother acceleration and idling performance. The better fuel consumption from the use of LPG is also attributable to the lack of cold–running enrichment (except for a few seconds during start–up only) that is necessary for petrol engines.

Where LPG is the sole fuel, experience has shown a number of advantages over petrol in respect of durability. Engine life is claimed to be typically 50 percent longer as a result of reduced cylinder bore wear during cold starting, since LPG does not wash oil off the cylinder walls and the lubricating oil has a longer effective life due to the almost total absence of dilution. Combustion chamber and spark plug deposits are reduced, although spark plug life is not necessarily extended. The exhaust system durability is extended when operating solely on LPG fuel (Menrad *et al*, 1985).

There is a penalty in power output resulting from the use of LPG instead of petrol. This is because of the amount of air the engine can induce – the gaseous fuel displaces more volume than petrol, metered as a liquid, and because in a petrol engine the latent heat of vaporisation of the fuel causes an increase in charge density as the petrol evaporates in the inlet system and cools the incoming air. The stoichiometric air/fuel ratio (AFR) for propane is 24.1:1 and for butane is 31.9:1 by volume (15.6 and 15.4:1 by weight, respectively). At the slightly richer AFR necessary for maximum power, LPG will displace 3–5 percent of the air/fuel mixture, and the expected power loss will be of a similar value. If the LPG engine AFR is set slightly lean for best economy, the power loss penalty will be somewhat higher at between five and ten percent. A very lean mixture for minimising exhaust emissions will lead to a substantial power loss of 20–30 percent relative to petrol.

It may be possible to mitigate part of the power loss associated with the use of LPG by increasing the charge density by cooling the inlet air by removing manifold heating or by utilising the latent heat of the fuel (with a pressure reducing valve and heat exchanger) in order to increase the mass of air induced by the engine. Another option may be to improve volumetric efficiency by increasing the size of the induction tracts and inlet valves, or by increasing the inlet pressure with a turbocharger or supercharger.

Relative to diesel, fuel consumption with LPG as a dual–fuel (using a mixture of LPG and diesel) on a thermal basis is unchanged or worse, depending on operating characteristics. At full engine load with LPG–only fuelling, fuel consumption may be 10 percent higher than with diesel fuel. LPG storage in a dual fuel diesel system is about two–thirds higher than for diesel on a volume basis for equivalent operating distance.

The drawbacks of the dual fuel system are firstly that it may be more difficult to fit an optimum design of LPG system, and secondly the bulk and complication of two fuel systems. In particular the LPG tank is pressurised (about 200 psi or 14 bar) and is heavier and requires more space than a petrol tank, for equivalent vehicle operating range. On a volume basis, the LPG fuel required is 15–20 percent more than petrol. LPG/air mixtures ideally require a considerable increase in spark plug energy compared with petrol/air mixtures. An improved ignition system, for example a higher voltage coil – by some 30–40 percent (Automotive Engineer, 1991a), or a capacitive discharge system, is therefore desirable. It is also desirable to optimise the ignition advance characteristics for use on LPG, but this may not be feasible on dual fuel engines. In addition, spark plug temperatures rise somewhat on LPG, so a change to a colder sparking plug may be necessary.

The starting load on the battery for an LPG engine tends to be higher than that for a petrol unit, due to the higher ignition system energy required and partly since the oil viscosity never becomes reduced by fuel dilution. Adequate fuel vapour pressure is necessary for cold starting and may be achieved by specifying a suitable minimum propane content for the prevailing ambient temperatures (propane has a higher vapour pressure than butane) or by providing electric pre–heating equipment.

6.2.1 Safety implications

Two potential safety and handling problems need to be emphasised with the use of LPG. When filling, it is necessary to maintain at least 10 percent free space for expansion in the LPG tank, otherwise thermal expansion of the liquid at high ambient temperatures could cause the safety valve to open and fuel to escape. The LPG industry is moving towards the adoption of automatic stop fill valves in the vehicle fuel tank (Webb and Delmas, 1991).

Both propane and butane are denser than air and tend to accumulate in low stagnant areas, producing a potentially serious fire hazard. Safety valves should discharge to atmosphere, not into enclosed spaces and it is common practice to provide vents for the vehicle compartment in which the LPG tank is mounted (for example, the boot) in case of leaks. LPG has similar flammability limits to petrol.

6.3   Feedstocks

LPG is derived from the lighter hydrocarbon fractions produced during petroleum refining of crude oil, and from the heavier components of natural gas which are removed before the gas is distributed. LPG represents about three percent of natural gas and a similar proportion of crude oil reserves although the production can vary by several percent depending on the feedstock and oil refinery design (Holman *et al*, 1991). In the United States about 60 percent of LPG is derived from natural gas processing plants where propane is separated from natural gas and the remaining 40 percent is produced from crude oil as byproducts during the refinery processing. The proportion of LPG available from natural gas may increase as supplies move from dry gas to other reserves and also because of increased gas use (ibid).

Large quantities of LPG are released and flared during natural gas and oil exploration and extraction. It has been estimated that, worldwide, some 150 million tonnes of LPG are flared, and that 70 million cars could be fuelled from this lost LPG alone (Hollemans and van Sloten, 1990).

Supplies of LPG in Europe and elsewhere are considered to be substantially expandable via increased gas plant recovery, increased refinery recovery, increased refinery production, the displacement of low value uses and the synthesis from gas and/or coal

(Webb and Delmas, 1991). It is expected that increased severity of refining processes will be required to meet future petrol octane requirements and to replace butane and aromatics (particularly benzene) to meet lower vapour pressure and aromatic toxics legislation. While increased refining severity will provide additional isobutane and olefin feedstock for conversion to alkylate for petrol, increased volumes of other LPG components will also become available (ibid).

6.4   Infrastructure

The transport and storage of LPG necessitates the use of pressure vessels at about 200 psi (1.4 MPa or 14 bar) to ensure the petroleum gas remains liquid at room temperature and adequate safety in handling the fuel are ensured by legal requirements and codes of practice. The specifications are usually defined for propane and are equally suitable for propane/butane mixtures since butane has a lower vapour pressure than propane. The use of pressure vessels has the advantage of fewer spillages and evaporation losses and contamination is reduced and the safety factor increased by virtue of the high standards set for the storage and transport vessels.

The storage, handling, transport and dispensing equipment, since the fuel is under pressure, is generally heavy, elaborate and expensive. On a volumetric basis, 15–20 percent more LPG would need to be transported for the petrol it displaces.

For LPG to be adopted on a larger scale as an alternative motor fuel, the delivery system will need to be expanded, particularly at the retail distribution level. Significant investments will be needed to provide service stations with multiple self–service LPG dispensers and buried storage tanks. However, the investment will be significantly lower than that required for CNG (reduced volume of fuel, stored at much lower pressure), or the even larger investment needed for methanol supply (Webb and Delmas, 1991). Research in New Zealand into the supply of LPG and CNG has shown (Kurani and Sperling, 1993) that only a relatively modest proportion (around 15 percent) of retail fuel outlets need to supply a new fuel to alleviate public concern about availability, even among non–users of the fuel.

6.5   Vehicle modifications

Engines developed by the engine manufacturer for the exclusive use of LPG is the preferred approach since the engine may be optimised for operation on LPG. This then requires the LPG storage tank and other fuel system components to be added.

6.5.1   Petrol–engined vehicles

Petrol engines may be converted to run on LPG, but maintain the original petrol fuel system and carburettor/fuel injection system. The drawbacks of the bi–fuel system are firstly that it may be more difficult to fit an optimum design of LPG system, and secondly the bulk and complication of two fuel systems. In particular the LPG tank requires more space than a petrol tank, for equivalent vehicle operating range.

There are three approaches to running spark ignition engines on LPG:

● Engines developed by the engine manufacturer for the exclusive use of LPG. This is the preferred approach since the engine may be optimised for operation on LPG. Features may include a higher compression ratio, larger bore induction system, improved valve and seat materials and an improved ignition system;

● Petrol engines converted to run exclusively on LPG. This is a possible option for fleet operators whose vehicles operate within a limited range of LPG outlets;

● Petrol engines converted to run on LPG, but maintaining the original petrol fuel system and carburettor/fuel injection system. The drawbacks of the bi-fuel system are firstly that it may be more difficult to fit an optimum design of LPG system, and secondly the bulk and complication of two fuel systems. In particular the LPG tank requires more space than a petrol tank, for equivalent vehicle operating range. Figure 13 shows a complete system for LPG bi-fuel operation, including a three-way catalyst and lambda sensor (from Automotive Engineer, 1991a).

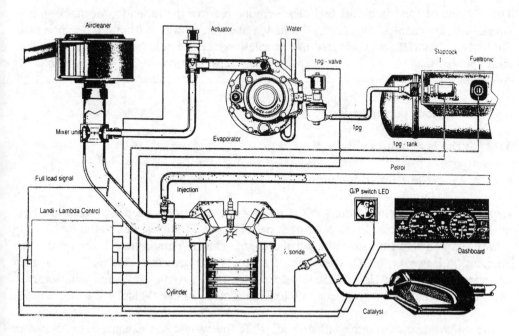

Figure 13.   A bi-fuel LPG system with three-way catalyst

The principle of operation of the LPG λ–control system, as shown in Figure 13, is that the control unit operates a stepper motor in the actuator to reduce the gas flow when a rich mixture is detected, and vice versa. The control logic differs for engines designed to run solely on LPG and those which can be run on petrol also. With fuel injected engines, there is an overlap between running on petrol and LPG, when the stepper motor progressively opens the actuator valve to increase the flow of gas before the petrol is cut off.

6.5.2 Diesel–engined vehicles

Diesel engines may be converted for operation on LPG as a secondary, or dual fuel. A disadvantage of the dual fuel approach, of course, is the duplication of much of the fuel system. Alternatively diesel engines may be converted to spark ignition to run on LPG alone, but a penalty in fuel consumption results and the conversion is more complicated and expensive than the conversion to run on LPG as a secondary fuel.

An example of dual–fuelling diesel–engined vehicles is in Vienna where many of the city buses are thus equipped, partially to take advantage of the low priced LPG. It is common to fuel with LPG (in addition to the diesel fuel) for higher engine power/loads only because of rough running, high CO and HC emissions and a drop in thermal efficiency at lower load operation.

The addition of LPG in a dual fuel diesel engine reduces the formation of black smoke, which usually enables the smoke–limited maximum power setting to be increased. Similarly a diesel engine converted to run solely on LPG would be expected to produce more power than with diesel because of better utilisation of the air available for combustion.

6.6   Emissions performance

6.6.1   Vehicle exhaust emissions

6.6.1.1   Substituted for petrol

Carbon dioxide emissions from LPG at point–of–use, for optimised engine technology, may be about 20 percent lower than from petrol–fuelled vehicles, due to the fuel having a lower carbon to hydrogen ratio and the engine able to demonstrate a higher thermal efficiency. However, a CO_2 reduction of about 11–12 percent seems realistic for cars equipped with a closed–loop emission–controlled LPG system, compared with a three-way catalyst–equipped petrol version (Hollemans and van der Weide, 1991).

Figure 14 overleaf shows the influence of AFR (by weight) on exhaust emissions with LPG (propane) as a vehicle fuel (from Shell, 1976). The characteristic curves are similar for both LPG and petrol with CO and HC emissions increasing at AFRs richer than stoichiometric and NO_x emissions peaking slightly lean of stoichiometric. Significant reductions of exhaust emissions are possible with LPG as a result of it being a gaseous fuel and able to be burnt at leaner AFRs than petrol. The largest reduction for an uncontrolled passenger car application is in CO emission, which has made LPG attractive for operation in confined spaces – for example, the operation of fork lift trucks in warehouses. Exhaust emission test results from the mid–1970s show CO emissions just 7½ percent of those from petrol cars, with HC emissions 55 percent and NO_x about the same (ibid).

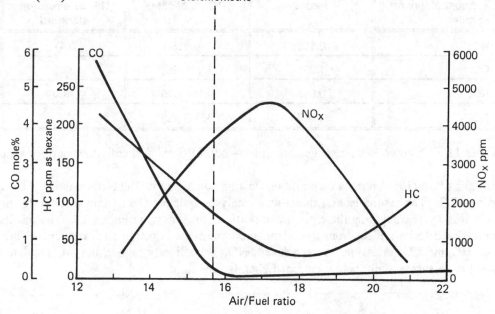

Figure 14. The influence of air/fuel ratio on exhaust emissions with LPG fuel

More recent research has been reported by Pitstick (1993) that reinforces that raw exhaust emissions of HCs and CO from LPG vehicles are lower than from petrol equivalents, but NO_x emissions have been about the same or higher. Modern LPG conversion vehicles with advanced emission control equipment have demonstrated very low emission rates, and dedicated vehicles that are developed specifically for LPG are expected to meet the California ULEV standards. Pitstick (1993) also reports that heavy–duty engine manufacturers are working on factory–produced LPG–fuelled engines with expected reductions in CO and HC emissions and very much lower particulate emissions.

Exhaust emission tests of a LPG (propane) dual–fuel conversion, closed–loop emission-controlled 1990–model Volvo 940 2.3–litre passenger car according to the US FTP 75 cycle have demonstrated very low emissions levels, as shown in Table 11 overleaf (from Hoogendoorn, 1991). The system was claimed to operate so efficiently that the exhaust emission of HC was less than the measured background reading (thereby giving a negative value). The vehicle was fitted with the Vialle Autogas Management System which controls the LPG/air mixture (it is also suitable for natural gas) to maintain λ=1. The system is suitable for both carburetted and fuel injected engines. It should be noted that these results were from the use of 100 percent propane, whereas commercially available LPG usually contains butane in addition, so a fairer comparison would have been obtained if tested with a representative mixture of propane and butane.

Exhaust emission (g/mile)	Petrol	LPG (% of petrol)	US '83 emission standard
HC	0.156	–0.018	0.41
CO	1.351	1.211 (90)	3.40
NO_x	0.595	0.136 (23)	1.00
CO_2	431.3	384.3 (89)	–

Table 11.   Volvo 940 passenger car closed–loop LPG exhaust emissions performance

Paramins Post (1993) reports on emissions testing conducted in The Netherlands by TNO using LPG. The results from a three–way catalyst–equipped Opel Omega, tested to the ECE R15 cycle (including the extra urban portion), are shown in Figure 15. The results are claimed to be typical from a modern, multipoint gas injection LPG system. While CO, HC and CO_2 emissions were reduced, NO_x was slightly higher, but all (regulated ones) were substantially below current EC emissions limits.

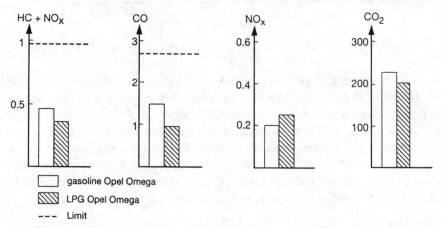

Figure 15.   Emissions from LPG–fuelled Opel Omega in the ECE cycle

6.6.1.2 Substituted for diesel

The operation of a dual fuel engine with diesel and LPG is claimed to reduce black smoke emissions significantly, accompanied usually by a reduction in CO but sometimes higher levels of HC and NO_x emissions. Conversion of a diesel engine to run *solely* on LPG is unlikely to affect CO and have little, or lower emissions of NO_x.

Experience at the TNO Road Vehicles Research Institute in the Netherlands with gaseous fuels for heavy–duty engines has shown that a stoichiometric LPG engine can meet the US HD 1994 emission standards and that a lean–burn version may meet the standard if a properly–adapted turbocharger, wastegate and intercooler and electronic fuel control system are used (van der Weide *et al*, 1988). Their results from emissions testing of LPG–fuelled lean–burn (turbocharged, intercooled) and stoichiometric (with three–way catalyst) heavy–duty engines according to the US heavy–duty diesel transient cycle are shown in Table 12.

Exhaust emission g/bhp-hr	Lean–burn LPG	Stoichiometric (3–w c) LPG	US HD Diesel 1994 standard
HC	0.5	1.05	1.30
CO	1.2	7.31	15.5
NO_x	4.0	0.9	5.0
Particulates	N/A	0.05	0.10

Table 12. LPG–fuelled heavy–duty engine exhaust emissions

It is believed that an important factor for the choice of fuelling strategy is the operation of the original engine. A petrol engine is today designed for stoichiometric operation while a diesel is designed as a lean–burn engine. In general it is advised to convert the engine for gaseous fuel to the fuelling operation for which it was originally designed (ibid), although this advice could prevent converted diesel engines achieving very low (NO_x) emissions.

6.6.1.3 Air toxics and secondary pollutants

A large fraction of the exhaust HC emissions from burning LPG is propane, which is less reactive than petrol and most other alternative fuel components, except methane. Propane is itself considered non–toxic and according to Pitstick (1993) it is likely that LPG vehicles will emit lower total rates of air toxics compared with petrol, with similar rates of formaldehyde and acetaldehyde emissions offset by much lower benzene and 1,3 butadiene.

Hollemans and van der Weide (1991), however, present emissions results and assert that aldehyde emissions from LPG–fuelled vehicles are lower than petrol equivalents, fitted with a similar emissions control system. Evaporative emissions from LPG vehicles are negligible, given the vapour–tight fuel systems, although they will have small refuelling losses during connection and removal of the supply line, although expected to be much lower than for petrol.

6.6.2 Life cycle emissions

Life cycle greenhouse gas (GHG) emissions are discussed in chapter 10 (**10.3.2**) with estimates ranging from a 17 percent reduction to a 12 percent increase, relative to petrol. Pitstick (1993) reports that DeLuchi (1991) has estimated GHG emissions from LPG (based on the current US LPG feedstock mix) to be 20 percent lower than petrol for light–duty vehicles and two percent lower for heavy–duty vehicles, compared with diesel fuel. Hollemans and van der Weide (1991) have estimated life cycle GHG emissions, taking account of recent LPG vehicle emissions test data in the Netherlands, to be some 5–6 percent lower for LPG than for petrol, assuming that the LPG is manufactured from natural gas. A decrease in the total GHG emissions would be realised if more flared natural gas was recovered for LPG production.

6.7 Costs

6.7.1 Fuel production and distribution

There tends to be only a limited market for LPG as a vehicle fuel, dependent on regional availability and tax incentives in certain countries, such as the Netherlands, Belgium, Italy, North America and New Zealand. In the Netherlands, while there is no excise tax on LPG sold as a vehicle fuel, there is an LPG surcharge on the vehicle road tax.

When LPG vehicles appeared in the UK in significant numbers in the 1960s the gaseous fuel was duty–free. In 1971 duty was imposed on gaseous fuels at the rate of 50 percent of that chargeable on leaded petrol, on a volumetric basis (a complex formula was devised to relate CNG to LPG so that the same rate of duty applied to both gases). With successive increases in the duty differential between leaded petrol and unleaded petrol and diesel, the economic incentive to switch to LPG has become eroded. Nevertheless, fuel savings for UK LPG vehicular use are claimed to be some 35 percent, and when reduced maintenance is taken into account, a 40 percent saving over petrol–fuelled vehicles is claimed (Tartarini, 1993).

A World Bank Technical Paper on alternative transport fuels suggests that for moderate quantities of LPG, shipment up to 300 km is usually by road in tankers with capacities of about 13,000 litres. The cost of distribution of LPG depends on the distance of the refuelling point from the source of supply, and the estimated percentage of distribution costs of the total fuel cost, are estimated (Moreno and Fallen Bailey, 1989) as follows:

Distance from source of supply	Distribution as % of total fuel cost
−5 km	2.2%
−50 km	7.4%
−100 km	11.1%
−300 km	23.5%

6.7.2 Vehicle modification

Of the various gas fuel systems manufactured in Italy, the USA and the Netherlands, perhaps the best known in the UK are those produced by Landi–Hartog. Autogas Installers and Retailers Association (the trade association for suppliers and installers of LPG and related equipment) have estimated the cost (in 1991) of a standard passenger car conversion to be £650. Tartarini (UK), a supplier of LPG conversion equipment, quote a conversion cost, including a 60–litre tank, in the region of £650 to £700 today.

New EC exhaust emission standards effectively requiring the fitting of three–way catalytic converters on petrol–engined cars will have an impact on the cost of LPG conversion to comply with the new emissions limits. The £650 quoted is likely to rise to perhaps £1,000, based on experience in the Netherlands. However Tartarini (1993) claim the cost of converting a catalyst–equipped fuel–injected petrol vehicle, such as a Volvo 760 GLE, to be in the region of £700 to £760, including a 70–litre LPG tank.

In the United States in 1991, the cost to convert a light–duty truck to LPG fuelling was estimated at between $800 and $1,500, while the cost to convert a passenger car may have been slightly higher, according to The Clean Fuels Report (1991b). A World Bank Technical Paper on alternative transport fuels estimates conversion costs for a range of alternatives, including LPG (Moreno and Fallen Bailey, 1989). Conversion costs for the following range of vehicles (in 1989) were:

- Passenger cars $700
- Light duty trucks $900
- Heavy duty goods vehicles $3,500
- Buses $4,800

6.8 Demonstration

There are currently about four million LPG vehicles operating worldwide with about 700,000 in the Netherlands, over 500,000 in Italy, about 400,000 in the USA and over 140,000 in Canada (Pitstick, 1993). Most European countries have an automotive LPG market, varying in size in relation to the excise duty chargeable on the fuel. Most vehicles using LPG are passenger cars (or light–duty vehicles) with petrol engines that have been retrofitted with commercially–available conversion components. The other major vehicle category for LPG fuel is the bus.

One of the largest urban bus fleets to operate on LPG fuel is in Vienna, where some 280 city buses were converted and have gained good driver and public acceptability due to low noise, vibration and smoke emissions. A closed–loop three–way catalyst and LPG control system has been developed by Deltec Fuel Systems and TNO that is suitable for fitment to increasing numbers of the Vienna bus conversions (van der Weide *et al*, 1991). More than 500 LPG buses were reported to be operating in Vienna in 1991 (The Clean Fuels Report, 1991a).

In New Zealand, the automotive LPG market grew rapidly during the 1980s, peaking in 1986 when over 14 million litres of LPG were sold for vehicle use. There are now about 50,000 LPG–fuelled vehicles operating, helped by lower excise taxes and a subsidised conversion programme sponsored by the conversion industry (ibid). Australia has about 200,000 vehicles running on LPG with some 1,600 retail outlets servicing these vehicles. In 1990 the conversions to LPG peaked at 42,000 that year.

Many American bus fleets have LPG conversions to evaluate the benefits from LPG operation. Other countries with major LPG fleets include Mexico, where an estimated 435,000 vehicles now use the fuel and Korea where 160,000 taxis are LPG–fuelled. In India, LPG is used as a diesel substitute to reduce expensive imports (Sathaye *et al*, 1988).

In the United Kingdom, the decline in cost–effectiveness of LPG as a motor fuel during the 1980s is reflected in the number of filling stations, which fell from over 500 in 1980 to around 200 or less today (Tartarini, 1993). However, operator interest in the fuel still exists, as shown by Stratford–based tour operator Guide Friday, who in 1993 has converted one of its Leyland Atlantean open–top buses to spark ignition LPG–fuelling, combined with a three–way catalytic converter. The move is a joint venture between the company and Danish–based Green Power International, which have together set up Green Power UK to undertake similar conversions for other operators. Green Power sees urban bus operation as an ideal application for LPG technology due to lower gaseous emissions and substantially lower engine noise. The fuel is supplied by British Gas.

6.9   Outlook

Liquefied petroleum gas has become an established fuel in many countries with a worldwide market penetration of about one percent of all motor fuels consumed. Several countries have launched major programmes to use LPG, for example in taxi fleets in Japan, Korea, Thailand, Spain and Greece, totalling over half a million vehicles already. Expectations are that the global LPG market will grow substantially in the coming years.

The former Soviet Union (now Commonwealth of Independent States – CIS), for example, intends to dramatically increase its use of LPG as a motor fuel (The Clean Fuels Report, 1991c). Having consumed 658,000 tonnes of LPG in 1990, the CIS is expected to increase consumption to five million tonnes by the year 2000.

Supplies of LPG in Europe and elsewhere are expandable via increased gas plant recovery, increased refinery recovery, increased refinery production, the displacement of low value uses and the synthesis from gas and/or coal (Webb and Delmas, 1991). The scale of expansion of LPG supplies in Europe, however, is uncertain since the largest growth in the use of LPG for automotive applications is expected to be outside of that market.

While emission benefits from the use of LPG with advanced emission control equipment have been shown in limited testing, the long–term emissions performance still needs to be assessed, relative to conventionally–fuelled vehicles. A major barrier to its widescale adoption is the cost of infrastructure and vehicle conversion, unless incentives, such as low or lower fuel taxes, are offered to persuade its greater usage.

In Europe, excise taxes have ranged from zero in the Netherlands and Belgium to 222 ECUs per 1,000 litres of LPG in Ireland, although Ireland's excise tax has been reduced to 100 ECUs per 1,000 litres leaving Denmark's tax the highest at about 160 ECUs per 1,000 litres of LPG (The Clean Fuels Report, 1991a). In the Netherlands, while there is no excise tax on LPG sold as a vehicle fuel, there is an LPG surcharge on the vehicle road tax. LPG currently has a 20 percent share of total Dutch motor fuel sales for passenger cars.

6.10   Summary

LPG is a mixture of propane and butane with a high octane rating (RON=112) for 100% propane. The fuel is liquid at room temperature under moderate pressure, but significantly less than the storage of compressed natural gas. It can be used as both a petrol and diesel fuel substitute.

<u>Advantages of LPG</u>

Good cold start performance, improved engine durability (no dilution of engine oil and reduced valve deposits) and exhaust system life extended.

Lower CO and HC relative to petrol and reduces formation of black smoke and particulates when substituted for diesel. Less reactive HC emissions and lower emissions of toxics (benzene and 1,3–butadiene). Negligible evap. emissions.

Dedicated LPG vehicles expected to meet California ULEV standard with ↓ HC (100%), ↓ CO (10%), ↓ NO_x (77%) demonstrated. ↓ CO_2 (11%).

Future heavy–duty emissions standards met without aftertreatment; significantly exceeded with.

Disadvantages of LPG

15–20% greater volume of LPG needed relative to petrol for equivalent range. Reduced power output of 5–10%, and up to 30% if fuelled very lean. Heavier tankage due to pressurisation.

REFERENCES FOR CHAPTER 6

AUTOMOTIVE ENGINEER (1991a). Special feature: Prospects for gaseous alternative fuels improved for LPG and LNG. Pages 38–41, February/March 1991.

AUTOMOTIVE ENGINEER (1991b). Viewpoint: Alternative fuels and power units. Pages 3 and 87, October/November 1991.

DELUCHI M A (1991). Emissions of greenhouse gases from the use of transportation fuels and electricity. Technical Memorandum No ANL/ESD/TM–22. Argonne National Laboratory, Argonne, Illinois, United States of America.

HOLLEMANS B and P VAN SLOTEN (1990). Perspectives of the automotive LPG market in Europe in 1995. TNO Road Vehicles Research Institute, Delft, The Netherlands.

HOLLEMANS B and J VAN DER WEIDE (1991). Use of LPG as an alternative fuel for light–duty vehicles: Its potential in relation to present and future emission demands. Proceedings of the symposium on the Use of Compressed Natural Gas (CNG), Liquefied Natural Gas (LNG) and Liquefied Petroleum Gas (LPG) as Fuel for Internal Combustion Engines, 23–27 September 1991, Kiev, Ukraine.

HOLMAN C, M FERGUSSON and C MITCHELL (1991). Road transport and air pollution: Future prospects. Rees Jeffreys Discussion Paper 25, Transport Studies Unit, Oxford University.

HOOGENDOORN J (1991). Alternative fuels – LPG. Proceedings from the seminar "Fuels, lubricants and the fleet operator", Institution of Mechanical Engineers, London, 4 December 1991.

KURANI K S and D SPERLING (1993). Fuel availability and dual fuel vehicles in New Zealand. Preprint paper No 930992, Transportation Research Board 72nd Annual Meeting, 10–14 January 1993, Washington, DC, United States of America.

MENRAD H, R WEGENER and H LOECK (1985). An LPG–optimised engine–vehicle design. SAE Technical Paper Series No 852071, Society of Automotive Engineers, Inc., Warrendale, Pennsylvania, United States of America.

MORENO R Jr and D G FALLEN BAILEY (1989). Alternative Transport Fuels from Natural Gas. World Bank Technical Paper number 98. Industry and Energy Series. Washington, DC, United States of America.

PITSTICK M E (1993). Emissions from ethanol and LPG fueled vehicles. Preprint paper No 930527, Transportation Research Board 72nd Annual Meeting, 10–14 January 1993, Washington, DC, United States of America.

SATHAYE J, B ATKINSON and S MEYERS (1988). Alternative fuels assessment: The international experience. International Energy Studies Group, Lawrence Berkeley Laboratory, University of California, Berkeley, California, United States of America.

SHELL INTERNATIONAL PETROLEUM COMPANY LIMITED (1976). LPG as an automotive fuel. Marketing–Oil Report 756, Fuels Development and Application – Motor Gasoline (alternative fuels), Shell Centre, London.

TARTARINI (UK) (1993). Private communication – 22 January 1993 – Mr G J Wise.

THE CLEAN FUELS REPORT (1991a). Propane (LPG): Industry analyses – BPN survey shows healthy status of international industry. Pages 162–163, November 1991.

THE CLEAN FUELS REPORT (1991b). Propane (LPG): Conoco teams with Firestone for LPG market. Page 159, November 1991.

THE CLEAN FUELS REPORT (1991c). Propane (LPG): Outlook and forecasts – World LPG forum sees increase in LPG use for motor fuel. Page 157, November 1991.

VAN DER WEIDE J, J J SEPPEN, J A N VAN LING and H J DEKKER (1988). Experiences with CNG and LPG operated heavy duty vehicles with emphasis on US HD diesel emission standards. SAE Technical Paper No 881657, Society of Automotive Engineers, Inc., Warrendale, Pennsylvania, United States of America.

VAN DER WEIDE J, R R J TER RELE and B HOLLEMANS (1991). Gaseous fuels in heavy duty engines: Developments from the Netherlands. Proceedings of the symposium on the Use of Compressed Natural Gas (CNG), Liquefied Natural Gas (LNG) and Liquefied Petroleum Gas (LPG) as Fuel for Internal Combustion Engines, 23–27 September 1991, Kiev, Ukraine.

WEBB R F and P J DELMAS (1991). New perspectives on auto propane as a mass–scale motor vehicle fuel. SAE Technical Paper No 911667, Society of Automotive Engineers, Inc., Warrendale, Pennsylvania, United States of America.

7. NATURAL GAS

7.1 Introduction

Natural gas (NG) has long been used in stationary engines, but the application of NG as a transport fuel has been considerably advanced over the last decade by the development of lightweight high–pressure storage cylinders. In addition to storage as compressed gas (CNG), natural gas can be liquefied (LNG) and stored cryogenically. It is estimated that in 1988 there were about 700,000 CNG–powered vehicles operating worldwide, and it is believed that in 1993 over a million vehicles in 25 countries were using natural gas.

Most existing natural gas vehicles (NGVs) use petrol engines, modified by after–market retrofit conversions and retain bi–fuel capability. Bi–fuelled vehicle conversions generally suffer from a power loss and can encounter driveability problems, due to the design and/or installation of the retrofit packages. Significant improvements in power and driveability can be realised with more sophisticated, vehicle–specific retrofit kits, or in factory–built bi–fuelled vehicles (US Congress, 1990). Engine and vehicle manufacturers are increasingly involved in the development of original equipment (OE), especially for heavy–duty (diesel replacement) applications, although many still regard the demand as too limited and dispersed to warrant large–scale manufacture.

Single–fuel vehicles optimised for natural gas are likely to be considerably more attractive in terms of performance, and somewhat more attractive in terms of cost. A natural gas–powered, single–fuel vehicle should be capable of similar power, similar or higher efficiency and mostly lower emissions than an equivalent petrol–fuelled vehicle. Such a vehicle would have a much shorter driving range unless the fuel tanks are made very large, which would then entail a further penalty in weight, space, performance and cost. CNG vehicles' range limitations, however, would be eased considerably if LNG were substituted as the fuel.

7.2 Fuel characteristics

Natural gas has a high octane rating (for pure methane, RON=130) enabling a dedicated engine to use a higher compression ratio to improve thermal efficiency by about 10 percent above that for a petrol engine, although it has been suggested that optimised NGVs should be up to 20 percent more efficient (DeLuchi et al, 1988a), although this has yet to be demonstrated. Optimum efficiency from natural gas is obtained when burnt in a lean mixture in the range $\lambda=1.3$ to 1.5, although this leads to a loss in power (which is maximised slightly rich of the stoichiometric air/gas mixture). Additionally, the use of natural gas improves engine warm–up efficiency, and together with improved engine thermal efficiency more than compensate for the fuel penalty caused by heavier storage tanks.

The use of natural gas as a vehicle fuel is claimed to provide several benefits to engine components and effectively reduce maintenance requirements. It does not mix with or dilute the lubricating oil and will not cause deposits in combustion chambers and on spark plugs to the extent that the use of petrol does, thereby generally extending the piston ring and spark plug life. In diesel dual–fuel operation evidence of reduced engine wear is reported, leading to expected longer engine life (Stephenson, 1991). The use of natural gas in a diesel spark–ignition (SI) conversion is expected to allow engine life at least as good as that of the original diesel engine.

Because of its very low energy density at atmospheric pressure and room temperature, natural gas must be compressed and stored on the vehicle at high pressure – typically 20 MPa (200 bar or 2,900 psi). The alternative storage method is in liquid form at a temperature of –162°C. Because of the limited capacity of most on–board CNG storage systems a typical gas–fuelled vehicle will need refuelling two to three times as often as a similar petrol or diesel–fuelled vehicle – a typical CNG–fuelled car will provide a range of 150–200 km and a truck or bus some 300–400 km. It is possible that the space required and weight of CNG fuel storage systems will fall in the future as a result of improved engine efficiencies (as with dedicated designs) and lightweight storage tanks.

CNG vehicles' range limitations would be eased considerably if LNG were substituted as the fuel. Rather than CNG's 4:1 volume disadvantage with petrol, LNG has only a 1.3:1 disadvantage. Even with the required insulation to ensure cryogenic storage, and the added bulk it causes, advanced LNG fuel tanks should be only about twice as bulky as petrol tanks storing the same energy (US Congress, 1990).

When a vehicle is operating on CNG about 10 percent of the induced airflow is replaced by gas which causes a corresponding fall in engine power output. In performance terms the converted bi–fuel engine will generally have a 15–20 percent maximum power reduction than that for the petrol version. When a diesel engine conversion is fuelled on gas *more* engine power can be obtained due to the excess air available which, due to smoke limitations, is not fully consumed.

Because natural gas has a low cetane rating a spark ignition conversion (for diesel engines) is required, adding to the conversion cost. Even though more power may be available, experience has shown that SI diesel engine conversions are usually downrated to prevent excessive combustion temperatures leading to component durability problems. A diesel/gas dual–fuel conversion may experience a loss of efficiency, relative to diesel–fuelling alone. A 15–20 percent loss in thermal efficiency was reported in a dual–fuel heavy–duty truck demonstration in Canada, where natural gas provided 60 percent of the total fuel requirement during dual–fuel operation (Sinclair and Haddon, 1991).

A further disadvantage of methane is that it is a greenhouse gas with a warming forcing factor many times that of the principal greenhouse gas, CO_2. Gas leakage or vehicular emission, therefore, and the size of release, will have an impact on the overall greenhouse gas (GHG) emissions performance relative to the petrol or diesel fuel it substitutes.

7.2.1 Safety implications

The safety aspects of converting vehicles to run on CNG are of concern to many people. However, the low density of methane coupled with a high auto–ignition temperature (540°C compared with 227–500°C for petrol and 257°C for diesel fuel) and higher flammability limits gives the gas a high dispersal rate and makes the likelihood of ignition in the event of a gas leak much less than for petrol or diesel (Stephenson, 1991). Additionally, natural gas is neither toxic, carcinogenic nor caustic.

The legal maximum operating pressure for a vehicle storage cylinder will usually be in the range 20 to 25 MPa – commonly 20 MPa. Cylinders are tested before installation to a pressure of 30 MPa (300 bar or 4,350 psi) or to a level to meet local requirements. Safety regulations specify a periodic re–inspection, typically at five–year intervals, including a pressure test and internal inspection for corrosion.

7.3 Feedstocks

Natural gas can be used as a fuel essentially in the form in which it is extracted. Some processing is carried out prior to the gas being distributed. Methane can also be produced from coal and from biomass (known as biogas) and a whole variety of biomass wastes (such as from landfill sites and sewage treatment plants).

Owing to the limited indigenous reserves, and because of the many uses for which it is favoured, it seems unlikely that UK–produced natural gas will find widespread use as a transport fuel (Holman *et al*, 1991). Nevertheless, worldwide natural gas reserves (as indicated in chapter 1) would appear to permit a large switch to it as an alternative transport fuel. Future UK natural gas demand is likely to be met from Eastern Europe and the CIS (former Soviet Union) imports.

7.4 Infrastructure

The use of natural gas as a vehicle fuel in the UK has the advantage of a comprehensive supply and distribution system already in place, thereby substantially reducing the cost of adopting it as an alternative fuel. A gas supply network has been in existence in the UK for around 150 years and is currently made up of 155,000 miles of distribution and transmission mains, connected to almost 18 million customers (Mitchell *et al*, 1990). However, the refuelling infrastructure would need to be established. In establishing a vehicular natural gas market the most attractive supply situation may be with fleet operators who have depots located in strategic locations around the country.

There are two refuelling modes with CNG:

- **fast fill** – where refuelling times are comparable to those involved with conventional liquid fuels. Fast fill normally requires some buffer high pressure (25 MPa) storage at the refuelling station although an alternative is to use a compressor sized to fill vehicles directly without intermediate (or cascade) storage. A typical medium–sized refuelling station with a compressor output of around 300 m³/hour would be capable of servicing 30 buses or 300 cars over a 12–hour period (Stephenson, 1991); and

- **slow fill (time fill or trickle charge)** – where one or more vehicles are connected directly to a low pressure supply via a compressor over relatively long time periods without the high pressure buffer storage facility. For many fleet operations the refuelling installation will be located at the fleet garage with trickle fill dispensers located adjacent to the vehicle parking spaces.

A CNG vehicle will be refuelled two to three times as often as a similar petrol or diesel counterpart. This has obvious implications for CNG refuelling station siting and local traffic flow constraints. The fact that gas is delivered by pipeline rather than by tanker, however, alleviates both traffic flow and road hazards. An additional consideration is the cost of connection to a gas pipeline having the pressure and flow capacity to meet the demand.

7.5   Vehicle modifications

The technology of engine and vehicle conversion is well established and suitable conversion equipment is readily available. An approximate measure of the equivalent petrol or diesel fuel capacity of a cylinder filled with gas at 20 MPa can be obtained by dividing the cylinder volume by 3.5 – thus a 60–litre cylinder will provide the energy equivalent of 17 litres of conventional fuel.

The design and installation of appropriate high–pressure on-board storage cylinders plays an important part of the efficient and safe operation of natural gas–fuelled vehicles. The cost constitutes a significant proportion of total vehicle installation cost. Most commonly used are chrome molybdenum steel gas cylinders, which are the cheapest, but one of the heaviest forms of storage container. It is possible that the space required and weight of CNG fuel storage systems will fall in the future as a result of improved engine efficiencies (as with dedicated designs) and lightweight storage tanks. For example, fibre–reinforced aluminium alloy or even all–composite CNG pressure tanks demonstrate significant weight saving over steel – up to 57 percent (Stephenson, 1991). It is even possible to increase the stored fuel's energy density by, for example, increasing the storage pressure of the gas. Future dedicated gas–fuelled vehicles will benefit by the fuel storage system being integrated into the vehicle structure, taking up less of the storage space currently lost in conversions.

One proposal for a future vehicle CNG storage system is the so-called "fortress frame". A modified vehicle frame structure, of significant cross-section, would be used to store the gas inside it at low pressure. Additionally, the frame would provide greater crash protection to the occupants (Int. J. of Vehicle Design, 1993). Although the design is likely to be as "safe" as conventional CNG vehicles, product liability issues, especially in the US, make the future development of this concept uncertain.

Research is in progress to use adsorbent materials in a tank to store natural gas which reduces the required pressure (from 200 bar for CNG currently, to around 30 bar) and thereby avoids the need for high pressure compressors and provides more design flexibility for the tank (IEA, 1990). Many types of adsorbent materials have been considered, including activated carbon, zeolites, clays and phosphates. With activated carbon at pressures of 300-400 psi (2-2¾ MPa or 20-27 bar), the percentage of natural gas adsorbed can be 10 to 15 percent of the weight of carbon. However, it has not yet been possible to find an adsorbent material which provides the same storage capacity of usable gas at the same cost, weight and volume as with high-pressure cylinders (ibid).

Although LNG storage has been used in demonstration fleets, few NGVs are operating on LNG at present. Advances are being made in local bulk LNG storage and, when vehicles are able to refuel their cryogenic storage tanks from such LNG depots at a cost that is competitive with CNG, more extensive use will be made of this form of storage. Until such time most vehicles using natural gas will store it in compressed form.

7.5.1 Petrol-engined vehicles

For spark ignition engines there are two options, a bi-fuel conversion or to use a dedicated, purpose-designed NG engine. The bi-fuel conversion of vehicles fitted with fuel-injected engines may utilise the original engine management system (if it can be modified to control the gas flow and revised ignition timing) or, alternatively, be fitted with a standard CNG control system. The fuel injectors must be disabled when the engine is running on gas, although fuel must still flow to the injectors (and then pass directly to the return fuel line) to provide cooling.

The two options for spark ignition (petrol) engines are:

- **bi-fuel engines** - spark ignition petrol engines of all sizes can be converted to natural gas by the fitting of a gas carburettor/mixer, regulator, shut-off valves, control system and fuel storage tanks. A bi-fuel arrangement exists when the petrol fuel system is retained, but this prevents the engine being fully optimised for the high-octane gas. This arrangement does provide a back-up fuel where CNG refuelling facilities are not well developed; and

- **dedicated natural gas engines** – dedicated natural gas engines are optimised for the natural gas fuel. They can be derived from petrol engines or may be specifically designed for the purpose.

Figure 16 (from British Gas, *undated*) shows the typical layout of CNG components in a petrol van conversion.

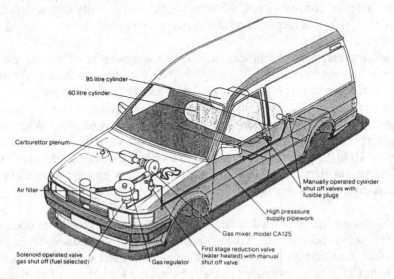

Figure 16.   Layout of CNG components in a petrol van conversion

7.5.2   Diesel–engined vehicles

With diesel–engined vehicles converted (or designed) to run on natural gas, there are two main options, discussed below:

- **dual–fuel engines** – these refer to diesel engines operating on a mixture of natural gas and diesel fuel. Natural gas has a low cetane rating and is not therefore suited to compression ignition, but if a pilot injection of diesel occurs within the gas/air mixture, normal ignition can be initiated. Between 50 and 75 percent of usual diesel consumption can be replaced by gas when operating in this mode. The engine can also revert to 100 percent diesel operation; and

- **dedicated natural gas engines** – dedicated natural gas engines are optimised for the natural gas fuel. They can be derived from petrol engines or may be designed for the purpose. Until manufacturer original equipment (OE) engines are more readily available, however, the practice of converting diesel engines to spark ignition will continue, which involves the replacement of diesel fuelling equipment by a gas carburettor and the addition of an ignition system and spark plugs.

For buses and trucks larger and greater numbers of cylinders are used than for light–duty vehicles, although the design of the vehicle chassis and bodywork plays a crucial role in their effective (and, if necessary, unobtrusive) integration. For compression ignition engine conversions to spark ignition, the pistons must be modified to reduce the original compression ratio and a high–energy ignition system must be fitted.

Lucas has developed a CNG injector system, currently fitted to dual–fuel diesel engine conversions. The system is suitable for both CNG and LNG (providing it is first vaporised and pressurised) and is ideally suited to timed (sequential) port injection systems, but can also be used for single point and low pressure in–cylinder injection. Gas fuel injection provides greater precision to the timing and quantity of fuel provided, and is likely to be further developed and become increasingly used to provide better fuel economy and emissions performance.

7.6 Emissions performance

7.6.1 Vehicle exhaust emissions

Because methane is a much simpler hydrocarbon than those in petrol and diesel and mixes more uniformly with air, combustion is likely to be more complete in the time available and leads to inherently lower CO and non–methane HC emissions. Because a gas–fuelled engine does not require cold–start enrichment, emissions from "cold" engine operation are lower than with liquid fuels, and because gas systems are designed to be air–tight, evaporative emissions *should* be negligible.

The generally low sulphur content of natural gas (5–10 ppm, mainly from the odourant) removes one of the important sources of catalyst poisoning and, for example, should allow the replacement of platinum by the much more abundant palladium. Palladium also has the advantage of being very effective for methane oxidation (Gratch, 1991).

The magnitude and character of emissions from NGVs, like emissions from alcohol and LPG vehicles, will vary depending on the trade–offs made between performance, fuel economy, emissions and other factors. However, the simple physical composition of natural gas (predominantly methane – CH_4 with no carbon–to–carbon bonds) tends to make it a basically lower emission fuel, since the combustion process is less complex than for liquid hydrocarbon fuels and there is less likelihood of incomplete combustion of the higher hydrocarbons.

7.6.1.1 Substituted for petrol

Methane has a lower carbon to hydrogen ratio than petrol or diesel fuel, and therefore produces less carbon dioxide when burnt. On an equivalent energy basis, methane produces 73% of the CO_2 emitted from the combustion of petrol (see Table 25 in chapter

10). Because the use of methane can improve engine efficiency, on a per unit distance travelled basis, allowing for a 10 percent efficiency improvement, natural gas produces around two-thirds of the CO_2 of a petrol vehicle, in a dedicated or modified engine.

Natural gas is used most efficiently when burnt in a lean mixture. Uncontrolled emissions of CO are very low compared with petrol exhaust emissions, as are non-methane HC (NMHC) emissions, although total HC (THC) can be higher due to the unburnt methane. Methane is difficult to control by catalyst because it is less reactive than many other HCs. While lean-burning natural gas engines (using an oxidation catalyst if necessary) are able to equal or better the CO and NMHC emissions performance of catalyst-equipped petrol engines (DeLuchi *et al*, 1988a), NO_x emissions may be higher. This is especially so if higher compression ratios are employed and is also because natural gas has a lower flame speed than petrol and requires ignition timing advance. Special combustion chambers and cylinder crowns, ensuring maximum swirl and squish, will help to partly offset this disadvantage in dedicated engines. Bi-fuel engines, because they are not optimised for natural gas operation, generally emit lower uncontrolled levels of NO_x when operating on gas than with petrol.

To address the problem of NO_x emissions legislation compliance, natural gas can be burnt in a stoichiometric mixture and a three-way catalyst used, although, as with the petrol engine, this does not provide the most fuel efficient solution. Several researchers have commented that NO_x reduction in natural gas engine exhaust was less efficient and required a richer mixture than is necessary for petrol engines (Weaver, 1989). It is possible that lower concentrations of reactive HCs in the natural gas exhaust may contribute to this, and that catalyst formulation may require changes if further research shows this to be the case. A fuelling strategy to partly overcome the fuel economy/NO_x tradeoff is to operate lean ($\lambda=1.3$ to 1.5) at low engine power and as engine speed and load rises, revert to stoichiometric operation, in combination with a three-way catalyst. This principle has been adopted by Toyota with its lean-burn petrol combustion system, T-LCS, used in the new Carina ε passenger car.

One potential problem that could occur with stoichiometric natural gas operation is the maintenance of the correct air/fuel ratio (AFR). Ryan and Callaghan (1992) report that varying natural gas composition can significantly alter the stoichiometric AFR. Tests at the Southwest Research Institute in California with a number of differing natural gas blends revealed that with a fixed orifice metering system a change in gas composition can have significant effects on the AFR, which is known to have significant effects on the levels of controlled emissions. Further investigation of this topic is ongoing at SwRI.

A Congressional Research Service report for the United States Congress concludes that natural gas-fuelled light duty vehicles (dedicated or bi-fuelled) are expected to meet the United States HC and CO emissions standards down to ULEV, but at this time have not been able to meet the NO_x standards at the LEV and ULEV levels (0.2 g/mile), even at the low mileage when emission controls are at their best. It is asserted that they are not yet reliably assured of meeting the 1994 model year (MY) or TLEV NO_x in–use standards of 0.4 g/mile (US Congress, 1992). This conclusion is borne out by Van der Weide and Seppen (1992), who report on demonstrated emission levels achievable with CNG–fuelled cars, tested to FTP procedures in The Netherlands. While the CO and NMHC emissions are within ULEV limits, NO_x emissions in the region of 0.16–0.21 g/km are within the TLEV limit (0.25 g/km equivalent) but above the ULEV requirements of 0.125 g/km.

Experience in Europe with CNG as a vehicle fuel, especially in Italy, suggests that controlled NGVs will be able to meet EC passenger car emissions standards, and even emissions from uncontrolled natural gas cars are almost low enough for compliance with some further development (Gambino *et al*, 1992). Tables 13 and 14 show emissions performance of a CNG–fuelled passenger car tested over the ECE R15 cycle. Table 13 shows results from an uncontrolled lean–burn Alfa Romeo 1.2–litre 4–cylinder boxer engine fitted with a CNG conversion. Table 14 details the results from the testing of a three–way catalyst–equipped Alfa Romeo 1.7–litre 4–cylinder boxer engine fitted with a CNG conversion supplied by Landi Renzo SpA.

ECE R15.04 test	Exhaust emissions g/km			
	CO	HC	NO_x	CO_2
91/441/EEC limits	2.72	0.97		–
Petrol engine	1.81	1.11	0.52	91
	1.63 combined			
CNG engine	0.76	0.64	0.35	64
	0.99 combined			

Table 13.   Lean–burn CNG–fuelled passenger car emissions performance

ECE R15.04 test	Exhaust emissions g/km			
	CO	HC	NO_x	CO_2
91/441/EEC limits	2.72	0.97		–
Petrol engine	0.22	0.32	0.15	122
	0.47 combined			
CNG engine	0.13	0.61	0.06	89
	0.67 combined			

Table 14.   Three–way catalyst CNG–fuelled passenger car emissions performance

The lean–burn CNG car displayed the highest thermal efficiency (17.5 percent over the R15.04 test) versus 14.8 percent of the stoichiometrically–fuelled CNG car. The petrol models displayed thermal efficiencies of 13.1 and 11.7 percent, over the same test, respectively. CO_2 emissions were also presented which indicate the NGVs to emit about 70 percent of those of the petrol models. Regulated emissions, especially CO, are very low and the combined HC+NO_x limit is only marginally exceeded by the lean–burn engine. The HC+NO_x emissions from the controlled CNG engine are higher than from the controlled petrol engine, caused by higher HC (mostly methane) emissions.

In the US, a gaseous fuel injection (GFI) system has been developed by ORTECH International in order to meet OE engine manufacturers' future requirements and for retro–fitting petrol engines for gaseous fuel operation. The system comprises a gas regulator, metering valve, gas introduction nozzle and computer, together with temperature and pressure sensors and a selector switch for the bi–fuel conversion. Following extensive testing, the California Air Resources Board (CARB) has certified the GFI dedicated/bi–fuel CNG conversion system for use on 1992 and older model–year passenger cars, light–duty trucks, medium–duty vehicles and heavy–duty engines. Testing of converted bi–fuel petrol/CNG passenger cars (with conventional three–way catalysts) showed typical initial results of 0.07–0.11 g/mile NMHC, 3.3–4.3 g/mile CO and 0.25–0.34 g/mile NO_x. More recent CARB testing has revealed lower emissions of NMHC and NO_x to below the California TLEV standards.

7.6.1.2   Substituted for diesel

The inherent advantage of natural gas when substituted for diesel fuel (rather than used with it in dual–fuel operation) is that it produces very low levels of particulate emissions. CO and HC emissions can be variable. Of greater concern is NO_x control, which is more difficult for natural gas than alcohol in a diesel engine conversion. Another environmental benefit is a reduction in engine noise, with kerbside noise levels falling from 88.5 dBA to 83.5 dBA (at 2,500 rpm) in road tests conducted with a natural gas–fuelled SI diesel engine conversion bus (Raine *et al*, 1987).

Cummins introduced a new heavy–duty direct injection diesel engine for worldwide bus and truck markets, including the UK, in the mid–1980s – the 6–cylinder, 10–litre L10 series. Because of its popular bus application, Cummins selected this engine for development as a natural gas–fuelled powerplant with the goal of meeting the US 1993 EPA bus standards. The conversion was to spark ignition using a high energy capacitor discharge system and a modified piston used to lower the compression ratio (to 10.5:1) and provide a faster burn. Best thermal efficiency when operating on natural gas has been demonstrated at 38 percent. A series of EPA heavy–duty diesel transient emission tests have been conducted and the range of recent emission values are shown overleaf in Table 15 (from Duggal, 1992).

US heavy-duty transient test procedure	Exhaust emissions g/bhp-hr			
	HC	CO	NO$_x$	Particulates
1991 California (1993 Federal) urban bus standard	1.30	15.5	5.0	0.10
Engine-out emissions	4-6	6-8	2.5-3.5	0.03-0.06
Oxidation catalyst-controlled emissions	0.6-0.9	0.3-0.6	2.5-3.5	0.03-0.05

Table 15.   Cummins L10 lean-burn natural gas engine emissions performance

Lower levels of NO$_x$ from the lean-burn L10 natural gas engine correspond with higher total HC levels, and vice versa. Thus NO$_x$ and HC tradeoff is evident, and that leaning of the mixture lowers NO$_x$ formation at the cost of increased THC. The experimental catalysts used for methane oxidation showed an 85 percent conversion efficiency (ibid). The results show that Cummins' design goals to meet 1993 Federal bus emissions standards are achievable. Commercial Motor (1992b) reports that the L10 natural gas engine is currently in operation in 350 buses in North America, although the power output of 179 kW is 63 kW less powerful than the L10 diesel version.

Development by Iveco of a 9.5-litre CNG-fuelled SI diesel engine conversion, combined with a three-way catalyst, is described by Cornetti *et al* (1992). Fiat Group has had over 20 years experience of natural gas as an automotive fuel, and after the second oil crisis, in the early 1980s, a 6-cylinder 9.5-litre gas engine (for CNG and LPG) was developed from the 8220.02 diesel engine.

A gas-powered bus fleet trial with the Iveco gas engine has taken place in Canada (10 LPG and 6 CNG buses) since 1984. The LPG trial finished in 1988, but the CNG project still continues, having covered over a million kilometres of bus operation as of June 1991. Due to NO$_x$ emissions from the lean-burn (for economy) CNG engines being about 60-90 percent higher than the diesel-fuelled equivalent, a decision was taken to develop the 8469.21 stoichiometrically-fuelled CNG engine, derived from the 8460.21 Iveco diesel engine. It comprises a spark ignition system, modified Heron-type pistons and reduced compression ratio (10.0:1), a turbocharger with electronically-controlled wastegate, closed-loop stoichiometric mixture control and a three-way catalytic converter. Exhaust emissions measured according to the ECE R49 13-mode cycle are shown overleaf in Table 16 (from Cornetti *et al,* 1992).

The Iveco CNG engine's uncontrolled NO$_x$ emissions were lower than the Canadian lean-burn engine developed during the 1980s, although are considerably higher than those from a diesel-fuelled version. Nevertheless, the catalyst significantly reduces pollutant emissions by up to 98 percent - although this is a new catalyst. Ageing can reduce a three–way catalyst's efficiency to 75 percent for heavy–duty utilisation, but with CNG it is expected to be no lower than 85 percent due to a reduction in lubricating oil consumption and consequent catalyst poisoning.

Emission limits according to EC Directive 91/542/EEC (Stages I and II) when tested to ECE R49 13–mode cycle	Exhaust emissions g/kWh			
	HC	CO	NOₓ	Particulates
Limits from July 1992 (Stage I)	1.1	4.5	8.0	0.36 †
Limits from Oct. 1995 (Stage II)	1.1	4.0	7.0	0.15 ‡
Engine-out emissions	2.1	13.7	15.1	–
Three-way catalyst-controlled emissions	0.52	0.87	0.96	0.012

† For engines ≤85 kW, limit is 0.61 g/kWh   ‡ Stage II particulate limit to be finalised

Table 16.   Iveco 8469.21 three–way catalyst–equipped natural gas engine emissions performance

7.6.1.3   Air toxics and secondary pollutants

Natural gas vehicles emit aldehydes, primarily in the form of formaldehyde. Limited testing of natural gas–fuelled vehicles has indicated that uncontrolled levels of formaldehyde emissions are considerably lower than those from methanol–fuelled vehicles and about comparable with those from uncontrolled emissions from petrol engines (US Congress, 1990), leading to the tentative conclusion (DeLuchi *et al*, 1988a) that formaldehyde emissions from catalyst–controlled NGVs should also be comparable with those from controlled petrol engines.

Because methane, the primary constituent of HC exhaust from a NGV is 100 times less reactive than methanol (the primary constituent of HC exhaust from a methanol–fuelled vehicle), the use of natural gas is expected to result in lower ozone formation than with the use of methanol as an alternative fuel to petrol (ibid).

7.6.2 Life cycle emissions

Life cycle fuel analyses of the use of natural gas show an expected reduction in GHG emissions relative to petrol. Several studies have estimated these greenhouse gas (GHG) emissions as CO_2 equivalent, and are shown in Figure 23 in chapter 10. The IEA (1990) quoting DeLuchi (1988b) show that CNG demonstrates a 19 percent advantage over petrol per vehicle mile driven. The use of LNG has been estimated at providing a 15 percent advantage over petrol, due to increased emissions from the use of energy to liquefy and distribute LNG (ibid). DeLuchi's analysis is moderately sensitive to the value of methane's global warming potential (GWP) and the amount of gas emitted from vehicles and lost during distribution.

Other studies show a range of estimates of greenhouse gas emissions from 27 percent better than from petrol, to 34 percent worse. Some studies made various assumptions that may not be currently accurate. The consensus appears to be that the use of CNG displays a fuel cycle GHG emissions advantage of around 20 percent over the use of petrol as an automotive fuel and is probably GHG emissions neutral compared with the use of diesel fuel.

A study of leakage from the UK natural gas distribution system (Mitchell *et al,* 1990) has estimated that a significantly higher percentage of supply than the 1 percent claimed by British Gas, is lost through fractured pre–1970 cast iron mains and old lead–yarn joints unable to maintain proper sealing as a result of dry natural gas replacing the town gas they were designed for. Additionally, mechanical joints kept tight by aromatic hydrocarbons (in town gas) had their sealing function reduced since natural gas lacks aromatic HCs. If the true leakage of natural gas from the UK distribution system is between 5.3 and 10.8 percent (the medium and high cases considered in the study) of supply, as the study authors think likely, this has significant implications for the global warming impact from the use of natural gas–fuelled vehicles.

7.7 Costs

7.7.1 Fuel production and distribution

Figure 26 in chapter 10 shows the IEA estimates of alternative fuel overall costs, relative to a baseline for petrol from crude oil, based on 1987 economics and fuel refining or feedstock conversion processes. The estimate of CNG overall costs range from 74 percent of the cost of petrol to 1.7 times its cost. It is possible, therefore, that natural gas–fuelled vehicles, especially fleets that bulk purchase their fuel, may be competitive economically with conventionally–fuelled ones. Most important are uncertainties as to future oil and gas prices. Any future tax imposed on the most polluting fuels (such as a "carbon tax") would favour natural gas in relation to oil, thereby assisting cost competitiveness.

Storage and refuelling stations represent a major cost item in the CNG fuel system for vehicles. A World Bank Technical Paper suggests that capital and operating costs for refuelling stations are more significant than feedstock cost in the ultimate delivered fuel cost. Moreno and Fallen Bailey (1989) calculated that for a typical United States–based CNG filling station having a throughput of 50 Mcf (1,416 m^3) per day, providing fast fill refuelling for about 200–250 light–duty vehicles, capital cost would be $150,000. Based on a natural gas cost of $1.00 per Mcf (thousand cubic feet) the delivered cost of $4.95 would comprise of $2.55 capital cost (annuitised over five years), representing 51.5%, $1.40 operating cost (28.5%) and the $1.00 feedstock cost representing 20% of the total delivered cost. While these costs are not necessarily representative for UK natural gas operation, they do illustrate the high proportion of refuelling station capital and operating costs.

The lowest cost supply would be a centralised vehicle fleet using a trickle fill system where the capital and operating cost would represent about 70 percent of the total delivered cost, estimated in the US at $3.30 per Mcf (ibid). Trickle fill refuelling systems are currently quite expensive, costing upwards of $1,000 per station, although mass–production should lower the costs (US Congress, 1990).

7.7.2 Vehicle modification

The conversion of a light–duty vehicle to bi–fuel operation is relatively straightforward, especially where vehicle and engine–specific conversion kits are commercially available. Stephenson (1991) estimates the cost for such a conversion to be between $600 and $1,200, depending partly on whether one or two gas cylinders are specified. The work required to convert a bus from diesel to SI natural gas operation is quite extensive. A practical example is the conversion of a MAN SL 200 (city bus), fitted with four 120–litre composite gas cylinders. Engine equipment cost $4,640, gas cylinders cost $2,400 and 182 hours of labour was needed (Stephenson, 1991).

A World Bank Technical Paper on the economics of alternative transport fuels estimates conversion costs for a range of alternatives, including those for natural gas (Moreno and Fallen Bailey, 1989). Conversion costs for the following range of vehicles to CNG in 1989 were:

- Passenger cars $1,000
- Light duty trucks $1,800
- Heavy duty goods vehicles $4,500
- Buses $4,500

Van der Weide *et al* (1991) indicate the cost increase of advanced, lightweight storage cylinders for CNG instead of conventional steel. The prices, relative to steel (100) and weight, relative to steel (100) per litre storage are: steel/kevlar – cost 210, weight 80; aluminium/composite – cost 180, weight 80; thermoplastic/glass fibre – cost 200, weight 50; thermoplastic/carbon fibre – cost 440, weight 30.

Stephenson (1991) presents a payback analysis, based on CNG conversion of a petrol–fuelled passenger car and a diesel bus, using the IEA energy prices published during the last quarter of 1989. The payback on a passenger car conversion costing $1,000 and using 2,000 litres of CNG per annum ranges from 0.7 years in Italy, 1.1 in The Netherlands, 1.2 in France, 1.9 in Belgium, 2.5 in Canada, 7.4 in New Zealand to 39 years in Australia. In the USA there is no payback period due to very low petrol prices. For a diesel bus conversion costing $10,000 and using 18,000 litres of gas per annum, the payback period ranges from 1.7 years in Italy, 1.9 in France, 2.2 in Canada, 2.7 in Belgium and Australia,

2.8 in The Netherlands, 3.1 in New Zealand to 5.6 years in the US.

In terms of payback period, New Zealand rates sixth (for petrol) and seventh (for diesel) out of the eight countries considered which may appear surprising due to the extensive vehicular use of CNG in that country. This situation can only be partly explained by the momentum generated in the industry during the 1980s when oil prices were high and CNG incentives favourable. Incentives covering part of the vehicle cost have been provided by the gas utilities and conversion suppliers, and are one reason for the continuing interest. The long payback period for petrol substitution in Australia is due to low petrol prices.

Dedicated, OE NGVs are expected to cost more than their petrol–fuelled counterparts because of the increased cost of the storage cylinders, pressure regulators and other components. For a light–duty vehicle, the US Department of Energy has estimated the cost increase to be $800, at 1990 US prices (US Congress, 1990).

7.8   Demonstration

It is reported that in 1988 there were about 700,000 CNG–powered vehicles operating worldwide, mostly in Italy with some 300,000 vehicles and 230 refuelling stations, New Zealand (130,000 vehicles and 400 filling stations) and Australia with over 100,000 vehicles (US Congress, 1990). Other countries with moderate numbers include the United States of America (30,000) and Canada (15,000).

Today it is believed that over a million vehicles in 25 countries are using natural gas as a fuel, with considerable CNG conversion programmes having also taken place in Argentina (135,000 vehicles), Indonesia (100,000), Thailand (50,000) and Pakistan with 21,000 vehicles included (IEA, 1990). Additionally, large–scale introduction plans for the former Soviet Union (now CIS) beginning in the early 1980s include 2,150 refuelling stations due to be operating at the end of 1990 (IANGV, 1990).

New Zealand's alternative fuels effort began in 1979, prompted by a large increase in the country's oil import costs. The Liquid Fuels Trust Board (LFTB) was formed with the target of substituting CNG, LPG and synthetic petrol for 50 percent of total demand. By 1986, total petrol demand was displaced 35 percent by synthetic petrol, 10 percent by CNG and three percent by LPG. Considerable market penetration has been achieved and promotion, initially a joint government/industry effort, is now primarily in the hands of the private sector with the target of 200,000 CNG vehicle conversions by the end of 1990.

A reduction in government grant support led to a fall in conversions in the mid–1980s and with a fall in petrol prices, the gas industry has struggled to remain competitive (Sathaye *et al*, 1988). Despite most New Zealand natural gas–fuelled vehicles being petrol conversions, there has been some diesel engine conversion activity. Auckland Regional

Council operates over 500 city buses and began CNG trials in 1987 with two conversions. As of August 1990 this had been extended to 35 buses and compressors being installed at other depots indicated a growing CNG bus fleet of about 20 per month (Stephenson, 1991).

Interest in the use of natural gas in **Canada** began to grow in the late 1970s, prompted by favourable experience in other countries and the growing price differential between natural gas and petrol, and Canadian LPG supply (Canada's first widely–used alternative fuel) was recognised as limited. Initial conversion costs were far higher than for LPG and a government grant programme for vehicle conversions and refuelling stations was launched in 1983. Many gas utilities provided financial incentives and leasing programmes to encourage vehicle conversions and refuelling station construction. Despite incentives, the conversion rate has been slow and payback periods for natural gas conversions have risen, partly due to lower petrol prices (Sathaye *et al,* 1988).

In the **United States**, General Motors Truck Division announced in July 1991 a contract to supply 1,000 natural gas pickup trucks, using composite wrapped aluminium storage cylinders, to a consortium of gas utilities, beginning in early 1993 (Automotive News, 1991). Initially converted by a specialist, the GM pickup trucks are expected to be eventually manufactured as dedicated vehicles. Ford also announced that it was to produce natural gas–powered pickup trucks for fleet trials during 1992. Chrysler plans to produce gas–fuelled vans that will also participate in fleet trials with a US Department of Energy–sponsored programme. By 1994 Chrysler plans to build production CNG vehicles (ibid).

In **Europe**, the Central Netherlands Transport Company has been a pioneer in natural gas–powered bus operation (Cragg, 1992). The transit authorities of Copenhagen, Helsinki, Oslo, Stockholm and Malmo have funded the Co–Nordic NG Bus Project. In **Sweden**, a trial of 20 CNG Volvo buses has been underway in Gothenburg. The single–deck buses, based on the conventional mid–engine B10M chassis, each have 15 CNG cylinders mounted on the roof, adding about 1,000 kg to the vehicle kerb weight and providing an operating range of 400–450 km. A Volvo B10B low–floor, rear–engine chassis could be used for mounting the gas cylinders underfloor.

In **France**, Gaz de France are collaborating with PSA in a demonstration trial of CNG–fuelled Citroën AX cars in Nantes. **Italy** has used CNG since the 1930s and was boosted after the 1973 oil crisis. Most CNG vehicles are private car conversions, primarily located in northern and central Italy where natural gas is produced. Italian use of CNG has never been actively promoted, but has instead resulted from high petrol prices (Sathaye *et al,* 1988). CNG bus conversion projects have been conducted in Italy – since 1986 Ravenna has been converting its bus fleet to CNG operation and decided to phase out all the diesels, replacing them more recently with buses specifically designed for CNG operation, using the Iveco 8460.21 engines (Montanari, 1992).

In the **United Kingdom** the use of CNG as a vehicle fuel is confined to a small number of road vehicles, mostly operated as part of a field trials project by British Gas. Since its acquisition of Consumers Gas of Canada in 1990, British Gas has expanded its NGV research (which began in 1988) in which it is investing over £1 million a year. British Gas research stations and several regions are participating in the trials, which involve bi-fuel conversions of small petrol vans (Maestro, Escort) and light commercial vehicles (Transit). Some 300 British Gas fleet vehicles were reported as running on natural gas in the autumn of 1992 (The Times, 1992). British Gas Scotland are investigating dual-fuel diesel conversions, and are now offering a conversion for light trucks, costing up to £4,000 (Commercial Motor, 1993). A 20 percent fuel saving over diesel-fuelled vehicles is claimed. Apart from refuelling facilities at five British Gas depots (with five more scheduled to open during 1993), the only other known natural gas filling station in the UK is in Blackburn where the Borough Council have been running converted light commercial vehicles in a fleet trial that began in 1986.

The UK's guiding body to coordinate all the threads of individual NGV progress and give direction to the development of the NGV market, is the Natural Gas Vehicle Association (NGVA). Members of the recently-established organisation include vehicle, engine and component manufacturers, service suppliers and fleet operators. The first objectives of NGVA include the determination of official standards for vehicle conversions and for the design and installation of refuelling stations and to initiate a programme to establish and promote a list of recommended suppliers, installation and service specialists.

7.9   Outlook

Natural gas is an attractive fuel that deserves careful consideration as an alternative to conventional fuels. Its use can offer reduced exhaust emissions, although with the introduction of catalytic converters on petrol-fuelled passenger cars the advantage may be small, except for carbon dioxide emissions that are typically 30 percent lower. Larger emissions benefits may be gained from heavy-duty diesel engine applications, but here the carbon dioxide emissions advantage is somewhat reduced. Its handling offers none of the toxicity problems of methanol or petrol and few of the explosion hazards of LPG, hydrogen, methanol and petrol. It appears not to pose a significant engineering challenge for vehicle and engine designers.

Several factors are likely to impede the rapid deployment of natural gas-fuelled vehicles across the general population. Bi-fuel vehicles may not perform as well as competing petrol vehicles (and certainly not in terms of range), so that the first generation of vehicle purchasers must be willing either to accept the limitations of these vehicles or to accept the risks of dedicated vehicles before an extensive infrastructure is in place. Further research, development and demonstration of the feasibility and benefits from the use of LNG are required, especially as several studies indicate that using LNG may be more economical than CNG in fleet applications (Pehrson, 1991, Sinor, 1992 and The Clean

Fuels Report, 1991).

Natural gas, in the first instance, is likely to find increased usage in vehicle fleets, such as buses and vans, where limited operating range is acceptable since the vehicles return daily to their depot for refuelling. The use of CNG in diesel engines is less established than for petrol engines, but may ultimately prove to become an attractive option, especially as cost benefits from using natural gas generally appear better for heavy–duty diesel conversion operation than for petrol since the extra weight of storage cylinders has a lesser effect on fuel economy than for cars. The achievement of low particulate emissions from urban buses (that generally have adequate under–floor gas storage space) is an example of such an application, and one which is likely to become more important in urban operation due to increased health concerns and increasingly stringent emissions legislation.

It was recently announced (Automotive News, 1993) that the "Big 3" US automakers (General Motors, Ford and Chrysler) have formed a research team, together with the natural gas industry, to accelerate the development and bringing–to–market of dedicated NGVs. The California ultra–low emissions requirement (from 1997) for light–duty vehicles is seen as one of the major factors in this recent collaboration. There is also rapidly growing interest in the US in the ability of CNG–fuelled engines to meet the HD particulate standards for buses in California since 1991 and the same Federal standard for buses in 1993 and trucks in 1994, and further reductions promulgated for future years.

An indication of the increasing importance of natural gas as a heavy–duty vehicle fuel has been the recent launch of several new engines. MAN launched a dedicated heavy–duty natural gas engine at the Brussels Commercial Motor Show in 1991. Based on the D 28 series 11.9–litre 6–cylinder diesel engine, it features spark ignition, modified compression ratio (12.5:1) and has closed–loop control in conjunction with a three–way catalytic converter. The E 2866 DF natural gas engine is able to meet Federal 1994 HD emission standards. Installed in the 19.232 tractor unit permits the fitting of eight 80–litre capacity CNG tanks – four each side of the chassis rails (MAN, 1991). MAN CNG engines have also been used, mostly as testbeds, in Australian city buses for a number of years.

Volvo has a 9.6–litre CNG bus engine that develops 185 kW – down 53 kW on the diesel equivalent (Commercial Motor, 1992b). The effective power loss is caused by the lean air/gas mixture and corresponding higher combustion temperatures than with diesel fuel for equivalent power output. To prevent component durability problems, therefore, an effective power downrating is necessary (as also highlighted with the Cummins L10 gas engine). DAF also plans to launch an 11.6–litre natural gas diesel engine which have been on trial in The Netherlands (Commercial Motor, 1992a).

In addition to Cummins' decision to launch a natural gas heavy duty diesel engine, Detroit Diesel introduced a 6V–92 dual–fuel bus engine in 1991, using an electronic fuel injection system developed for the company's methanol engine (Mechanical Engineering, 1991). More recently, Detroit Diesel has launched a dedicated NG engine for bus and truck applications. The Series 50G is a lean–burn 8.5–litre four–stroke engine derived from the company's Series 50 diesel unit, and has the same power and torque rating. The engine has been designed to meet the California transient emissions standards for heavy–duty natural gas engines (Automotive Engineering, 1994).

Worldwide natural gas usage is forecast to rise substantially in the coming decades. It is replacing coal for power generation and many other alternative fuels, should they become used more widely, use natural gas as a feedstock. The fuel oxygenate, MTBE, is such an example whose future production is forecast to increase dramatically. Whether, in say a decades' time, there will be sufficient *low–cost* gas to support its diversion to transport fuels on a scale large enough to significantly enhance energy security or help limit emissions on a global scale, is debatable. Natural gas, however, can still contribute to "cleaner" urban environments, at least in the medium term.

7.10   Summary

Natural gas is mostly methane, and has a high octane rating (RON=130). It can be used as both a petrol and diesel fuel substitute in either bi–fuel or dual–fuel conversions, or in dedicated NG engines with spark–assistance.

Advantages of NG

NG's high octane rating enables dedicated NG engines to become more efficient through higher compression ratios. If fuelled lean, an engine can display a significant economy improvement. Maintenance requirements are generally lower for a NG engine.

For petrol substitution, emissions performance depends on fuelling strategy: lean–burn leads to ↓ CO, variable HC (although ↓ NMHC), variable NO_x (depending on compression ratio), ↓ CO_2 (up to 30%). Using a three–way catalyst produces very similar, or slightly lower, emissions performance to petrol.

Main advantages for dedicated heavy–duty engines are very low particulates, although exact emission levels also depend upon fuelling: lean–burn leads to very low CO and very low particulates. A three–way catalyst causes low NO_x and even lower particulate levels.

Disadvantages of NG

NG causes a reduction in power by up to 20% in a bi–fuel engine. The fuel must be stored in compressed form at high pressures, liquefied at a low temperature or adsorbed by another material. All these imply increased weight, volume and cost over conventional fuel tanks.

REFERENCES FOR CHAPTER 7

AUTOMOTIVE ENGINEERING (1994). Tech Briefs: Heavy–duty natural gas engine. Page 17, Vol. 102, No. 2, February 1994.

AUTOMOTIVE NEWS (1991). Clean air rules renew interest in natural gas. Pages 16 and 18, 4 November 1991.

AUTOMOTIVE NEWS (1993). Big 3 and SAE team to spark research for natural gas use. Pages 1 and 35, 1 March 1993.

BRITISH GAS (Undated). Natural Gas Vehicles: The way ahead... naturally. British Gas pamphlet, British Gas, Staines.

COMMERCIAL MOTOR (1992a). News: Gas engine due for 1992 launch. Page 16, 20–26 February 1992.

COMMERCIAL MOTOR (1992b). News: Cleanliness gives gas edge. Page 14, 31 December 1992–6 January 1993.

COMMERCIAL MOTOR (1993). Business news: BG's 20% dual–fuel saver. Page 11, 20–26 May 1993.

CORNETTI G M, F FILIPPI and M SIGNER (1992). CNG low emission city buses. Proceedings of the International Conference "Bus '92. The Expanding Role of Buses Towards the Twenty–First Century", 17–19 March 1992, Institution of Mechanical Engineers, London. C437/030.

CRAGG C (1992). Cleaning up motor car pollution: New fuels and technology. Financial Times Management Report, FT Business Information Ltd., London.

DELUCHI M A, R A JOHNSTON and D SPERLING (1988a). Methanol vs. natural gas vehicles: A comparison of resource supply, performance, emissions, fuel storage, safety, costs, and transitions. SAE Technical Paper Series No 881656, Society of Automotive Engineers, Inc., Warrendale, Pennsylvania, United States of America.

DELUCHI M A, R A JOHNSTON and D SPERLING (1988b). Transportation fuels and the greenhouse effect. Transportation Research Record 1175, National Research Council, Washington, DC, United States of America.

DUGGAL V K (1992). The natural gas L10 urban bus engine – an alternative fuel option. Proceedings of the International Conference "Bus '92. The Expanding Role of Buses Towards the Twenty–First Century", 17–19 March 1992, Institution of Mechanical Engineers, London. C437/026.

GAMBINO M, S IANNACCONE and A UNICH (1992). Frontier of technologies: The dedicated engines. Pre–prints of proceedings from the THERMIE European Seminar "Natural Gas as Fuel in Public Transport Vehicles", 13–15 May 1992, Milan, Italy.

GRATCH S (1991). Survey of CNG vehicle technology. Report for Auto/Oil Air Quality Improvement Research Program, CRC Contract No. AQIRP–17–91.

HOLMAN C, M FERGUSSON and C MITCHELL (1991). Road transport and air pollution: Future prospects. Rees Jeffreys Discussion Paper 25, Transport Studies Unit, Oxford University.

IANGV (1990). A position paper on Natural gas vehicles 1990. International Association for Natural Gas Vehicles (ed. J Stephenson), Auckland, New Zealand.

INTERNATIONAL ENERGY AGENCY (1990). Substitute fuels for road transport: A technology assessment. OECD/IEA, Paris.

INTERNATIONAL JOURNAL OF VEHICLE DESIGN (1993). Technical Note: Fortress Frame CNG vehicle. Pages 390–395, Vol. 14, No. 4. Inderscience Enterprises Ltd., UK.

MAN (1991). Press Release: New at the Brussels Commercial Vehicle Salon 1991 – MAN truck 19.232 with 230 hp natural gas engine. MAN Nutzfahrzeuge Aktiengesellschaft, Munich, Germany.

MECHANICAL ENGINEERING (1991). Alternative fuels: Paving the way to energy independence. Pages 42–46, December 1991.

MITCHELL C, J SWEET and T JACKSON (1990). A study of leakage from the UK natural gas distribution system. Pages 809–818, Energy Policy, November 1990, Butterworth–Heinemann.

MONTANARI R (1992). Running public transport services with CNG fuelled buses. Pre–prints of proceedings from the THERMIE European Seminar "Natural Gas as Fuel in Public Transport Vehicles", 13–15 May 1992, Milan, Italy.

MORENO R Jr and D G FALLEN BAILEY (1989). Alternative Transport Fuels from Natural Gas. World Bank Technical Paper number 98. Industry and Energy Series. Washington, DC, United States of America.

OCCUPATIONAL SAFETY & HEALTH (1993). Road report: Stepping on the gas. Page 31, January 1993.

PEHRSON N C (1991). LNG vehicle demonstration projects. SAE Technical Paper Series No 911661, Society of Automotive Engineers, Inc., Warrendale, Pennsylvania, United States of America.

RAINE R R, S T ELDER and J STEPHENSON (1987). Optimisation of diesel engines converted to high compression spark ignition (SI) natural gas operation. Proceedings of the 4th International Pacific Conference on Automotive Engineering "Mobility: The technical challenge", 8–14 November 1987, Melbourne, Australia.

RYAN T W and T J CALLAGHAN (1992). The effects of natural gas composition on engine combustion, performance, and emissions. Proceedings from the XXIV FISITA Congress "Automotive Technology Serving Society", 7–11 June 1992, London.

SATHAYE J, B ATKINSON and S MEYERS (1988). Alternative fuels assessment: The international experience. International Energy Studies Group, Lawrence Berkeley Laboratory, University of California, Berkeley, California, United States of America.

SINCLAIR M S and J J HADDON (1991). Operation of a class 8 truck on natural gas/diesel. SAE Technical Paper Series No 911666, Society of Automotive Engineers, Inc., Warrendale, Pennsylvania, United States of America.

SINOR J E (1992). CNG/LNG issues and comparisons. J E Sinor Consultants Inc., Niwot, Colorado, United States of America.

STEPHENSON J (1991). Learning from experiences with Compressed Natural Gas as a Vehicle Fuel. Centre for the Analysis and Dissemination of Demonstrated Energy Technologies (CADDET) analyses series number 5. Sittard, The Netherlands.

THE CLEAN FUELS REPORT (1991). Norway considering LNG transportation systems: Cost benefits. Page 126, November 1991.

THE TIMES (1992). Motoring: Gas is naturally greener. Page 9, 25 September 1992.

US CONGRESS (1990). Replacing gasoline: Alternative fuels for light–duty vehicles. Office of Technology Assessment, US Government Printing Office, Washington, DC, United States of America.

US CONGRESS (1992). Alternative fuels for automobiles: Are they cleaner than gasoline? CRS Report for Congress 92–235 S, Congressional Research Service, Washington, DC, United States of America.

VAN DER WEIDE J, R R J TER RELE and B HOLLEMANS (1991). Gaseous fuels in heavy duty engines; Developments from The Netherlands. Proceedings from the UNECE Symposium on the Use of Compressed Natural Gas (CNG), Liquefied Natural Gas (LNG) and Liquefied Petroleum Gas (LPG) as Fuel for Internal Combustion Engines, 23–27 September 1991, Kiev, Ukraine.

VAN DER WEIDE J and J J SEPPEN (1992). Natural gas vehicles, environmental balances. Pre–prints of proceedings from the THERMIE European Seminar "Natural Gas as Fuel in Public Transport Vehicles", 13–15 May 1992, Milan, Italy.

WEAVER C S (1989). Natural gas vehicles – a review of the state of the art. SAE Technical Paper Series No 892133, Society of Automotive Engineers, Inc., Warrendale, Pennsylvania, United States of America.

8. ELECTRICITY

8.1 Introduction

The electric vehicle (EV) concept is more than 100 years old, the first vehicle appearing in 1890. It was in the 1960s that interest in EVs was revived, mostly due to air quality concerns in the USA. Developments continued slowly, boosted slightly by the oil crisis of 1973–74, but not always given a particularly high priority. Many vehicle manufacturers built prototypes, mostly in order to gain experience of electric traction systems rather than to consider marketing EVs at the time.

Failure of research and development efforts to produce the much hoped for breakthrough in EV battery technology was one of the reasons for declining interest in EVs after oil prices peaked in 1981 (DeLuchi *et al*, 1989). It was towards the end of the 1980s that EV development was really considered seriously as environmental issues became especially heightened due to deteriorating air quality in many urban areas worldwide.

The use of EVs, in adequate numbers, would be expected to confer excellent urban pollution benefits, since they emit no pollution at point–of–use. The infrastructure for electricity distribution is established (except for recharging facilities) and night–time excess generating capacity should be adequate to fuel a very substantial EV market. However, the impact of EV usage has to take account of the additional pollution emitted from the power stations that provide the electricity. Additional limitations, at the present time, include poor range and generally bulky, heavy batteries that make the EV uncompetitive with conventional vehicles. Very significant development effort is still required to transform the current EV into a major transport provider of the 21st century.

8.2 Energy storage

EVs use electricity that can be stored in batteries that are recharged from the mains supply or, alternatively, generate it on board the vehicle using a heat engine and generator or use fuel cells to provide the electricity. In both latter cases batteries would be used to provide an energy load–levelling system that allows vehicle performance that may demand more power than the engine can itself generate and also allow all–electric operation (a Zero Emission Vehicle – ZEV) if necessary.

The so–called hybrid vehicle does not need to rely exclusively on mains electricity for battery recharging. The reader is referred to the companion volume to this book, *Alternative Engines for Road Vehicles,* the research and review for which was carried out in parallel to, and in conjunction with, this one. One chapter in the Engines book is devoted to the review of hybrid vehicles, most of which are the internal combustion engine/battery type.

8.2.1 Battery technology

Early prototype EVs produced in the 1960s mostly used tubular lead–acid batteries. At the end of the 1960s Chloride began the development of sodium–sulphur batteries, as did some other companies in Europe. In addition, nickel–iron, nickel–zinc, iron–air and zinc–air batteries began to be developed. During the last decade battery technology has made substantial progress, but significant advances are still required, especially in terms of increasing the energy density – to provide range, and power density – to provide vehicle performance. High costs (frequently due to the limited or prototype nature of some of the batteries in question), questionable durability, short shelf life and uncertainties of the ability to recycle some of the sometimes toxic battery components relate, in some combination, to many of the battery technologies being proposed for future EVs.

A description of the various battery technologies being developed for EV applications are presented in the next section, after a summary table of their relative performance. It may be useful to compare present battery performance with the United States Advanced Battery Consortium (USABC) criteria for both mid and long–term acceptance. The USABC objectives are to develop an advanced battery to meet mid–term criteria with pilot plant prototype production in 1994 and demonstrate the design feasibility of an advanced battery to meet the long–term criteria in 1994. The mid–term criteria were defined to improve the batteries of EVs to be introduced in the mid–1990s, to effectively double the range of current EVs, and the long–term goals would allow EVs to have performance of, and be cost–competitive with, conventional vehicles by the late 1990s. Table 17 shows some of the USABC primary criteria.

Parameter	USABC primary criteria	
	Mid term	**Long term**
Energy density (Wh/kg) @ 3–hour rate	80 (100 desirable)	200
Power density (W/kg) @ 80% depth of discharge (30–sec)	150 (200 desirable)	400
Life (years)	5	10
Cycle life (cycles to 80% DOD)	600	1,000
Ultimate cost ($/kWh)	<150	<100
Operating environment temperature range	–30 to 65°C	–40 to 85°C
Recharge time (hours)	<6	3 to 6

Table 17.   USABC primary criteria for EV battery performance

The summary of the status of current battery technology is shown overleaf in Table 18. Where possible, data is provided from demonstrated battery testing over standard charge–

discharge cycles. Test results for some technologies at an advanced stage, where complete batteries have not been available for testing, are for individual cells. Comparison of these with data from battery testing is not necessarily reliable – some batteries suffering from a few individual cell failures have performed much worse than others, depending on their cell configuration. Additionally, scaling up the results from cell testing will not necessarily equate to the complete battery performance – for example, some require heating which uses energy, thereby reducing the overall performance of the battery.

Battery	Energy density[1] Wh/kg	Power density[2] W/kg	Operating temperature °C	Cycle life[3]	Production cost £/kWh	Commercial availability
Lead–acid (Pb–A)	30–35	100	Ambient	750–1000	35–50	Now
Advanced lead–acid (Pb–A)	35–40	125	Ambient	700+	90 [40]	1994–95
Nickel–cadmium (Ni–Cd)	55	190	Ambient	600+	300	Now
Nickel–iron (Ni–Fe)	55	110	Ambient	750+	120	1994–96
Sodium–sulphur (Na–S)	100	150	300–350	660+	[50–100]	1993–95
Sodium–nickel chloride (Na–NiCl$_2$)	80–100	100–110	300	700+	[75]	1994–96
Lithium aluminium–iron sulphide (LiAl–FeS)	_96_	_166_	400–465	_240+_	[60]	1995–2000
Lithium aluminium–iron disulphide (LiAl–FeS$_2$)	_180_	_400_	400	–	–	2000+
Lithium–solid polymer	_160_	_120_	80–120	300	[50–225]	2000+
Aluminium–air (Al–air)	[220]	–	Ambient	–	–	–
Zinc–air (Zn–air)	[90–200]	–	Ambient	[600+]	–	–
Zinc–bromine (Zn–Br)	80	55	Ambient	337	[50]	–
Nickel–metal hydride	_55_	_175_	Ambient	_333_	–	2000+

Notes:

[1] At the 3–hour discharge rate. Indicator of vehicle range
[2] At ≥50% depth of discharge (DOD). Indicator of vehicle performance
[3] Replacement when recharge capacity <80% of rated battery capacity
Data is for demonstrated battery technology as of 1990/92
[Anticipated performance is shown in parentheses thus]
Results from **cell** testing, not complete battery, are shown underlined

Table 18. EV battery technology status (from Braithwaite and Auxer, 1991, Burke, 1991, Cheiky *et al*, 1990, De Jonghe *et al*, 1991, DeLuca *et al*, 1989, DeLuca *et al*, 1991, Henriksen and Embrey, 1991, McLarnon and Cairns, 1989, New Scientist, 1992, Westbrook, 1992)

8.2.2 Lead–acid

The lead–acid ($Pb/H_2SO_4/PbO_2$) battery is the most widely used rechargeable electrochemical device. It is a mature technology, discovered about 100 years ago. Low-maintenance and maintenance–free batteries have been developed and better electrode materials have reduced corrosion problems, although this remains a life–limiting problem. Lead–acid batteries designed for EV applications can now deliver more than 40 Wh/kg – a 30 percent improvement over the performance available a decade ago (McLarnon and Cairns, 1989). Safety hazards posed by the sulphuric acid liquid electrolyte can be minimised by using it in gelled form, and the release of hydrogen during recharging can be contained in a sealed battery construction. The performance of Pb–A batteries falls at low temperatures.

Today's Pb–A batteries are expected to last for 30,000 miles in an EV with deep discharge cycles and longer with light use combined with frequent charging. Advanced, maintenance–free (MF) designs are expected to last for 1,000 discharge cycles, and provide up to 60,000 miles of EV use (EPRI, 1991). One of the most advanced MF systems (DRYFIT) is manufactured by Sonnenschein in Germany, and comprises flat-plate technology. The tubular plate design of Chloride Motive Power is incorporated in GM's electric G–van, and while not presently maintenance–free, it is expected to become so in the future.

8.2.3 Nickel–cadmium

Most interest in Ni–Cd battery technology is in Europe and Japan. It offers 1½–2 times the vehicle range to that provided by Pb–A and good vehicle performance characteristics due to its high peak power capacity, even at high rates of discharge. Costs are currently very high compared with other battery technologies, although these are anticipated to fall as EV Ni–Cd battery technology advances. However, nickel–metal hydride batteries are now thought to be the Ni–Cd future replacement, so production costs for Ni–Cd may not fall as low as earlier expected if the technology is largely replaced (at least in EV applications) in a decade's time. Additionally, there have been concerns expressed about the toxicity of cadmium and its worldwide availability should EV applications require substantial production increases. Ni–Cd battery manufacturers claim that recycling of the cadmium from Ni–Cd batteries can be 100 percent and that collection and recycling schemes are being established.

The French battery manufacturer, SAFT, is a leading developer of alkaline battery technology and for the past decade has been working with vehicle manufacturers on applications for Ni–Cd batteries in EVs. Today, many commercially available EVs are fitted with Ni–Cd batteries to provide greater range than from Pb–A, but of course at higher cost. The vented–plate design are mostly used in EV applications, since the more advanced sealed cells are very expensive.

8.2.4 Nickel–iron

The nickel–iron battery is also known as the iron–nickel oxide (Fe/KOH/NiOOH) battery or Edison cell. It exhibits greater specific energy than the lead–acid battery and is known for long life and ruggedness. Both the Fe and NiOOH electrodes are inherently inefficient and the battery's energy efficiency is typically about 60 percent (McLarnon and Cairns, 1989). For EV applications, this inefficiency leads to the need for frequent water addition (topping up). The hydrogen evolved at the Fe electrode during recharging may cause safety problems. Ni–Fe batteries, expected to be commercially available in 1994–96, can provide 1½–2 times the vehicle range of Pb–A, but at increased cost.

The Ni–Fe battery has been extensively tested in the TE–Van programme at Chrysler, and is still considered to be a viable candidate. Extensive development work and testing was carried out in Europe, notably by SAFT and Peugeot, but was discontinued in about 1990, mostly due to iron poisoning problems caused by impurities in the electrolyte.

8.2.5 Sodium–sulphur

Batteries that operate at elevated temperatures exhibit improved performance compared with ambient–temperature batteries. However, it is necessary to insulate them to prevent rapid heat loss, and provide a heat source which reduces the battery performance. The sodium–sulphur (Na/S) or BETA battery has been under development for more than 20 years. It uses a molten Na negative electrode, a solid Na^+ ion conducting electrolyte and a molten sulphur–sodium polysulphide mixture as the positive electrode. Elevated temperatures of typically 300–350°C are required to achieve sufficiently high ionic conductivity of the solid electrolyte.

Sodium–sulphur cells being developed for electric vehicles have demonstrated energy densities as high as 165 Wh/kg and power densities of 220 W/kg, but *battery* performance is more modest, owing in part to the need to provide a thermally insulating enclosure with heating provided from the battery's own stored energy (ibid). Premature cell failures have proved a problem and can complicate cell interconnection strategies because cells may fail either shorted or open circuit, and because individual cells will fail if subjected to overcharge or discharge. Advantages include the potential for low-cost, long life (1000–2000 deep discharge cycles) and high efficiency (overall, including heat loss of around 176 W for a 22 kWh battery, can be up to 90 percent). Remaining issues to be resolved include the improvement in service life (currently just over a year), improvement in the thermal insulation and the identification of effective battery reclamation techniques (Braithwaite and Auxer, 1991).

Some commentators believe that the Na–S battery may provide the best hope of providing EVs with adequate range at acceptable cost by the end of this decade (EPRI, 1991). Over the last two decades the two major developers of this technology have been Chloride

Silent Power in the UK and the Swedish–Swiss company ABB Asea Brown Boveri. ABB has a pilot plant in Germany and Chloride has formed a joint venture with the German company RWE for a pilot plant in the UK to manufacture Na–S batteries. ABB has begun series production, initially around 500 units annually, but expects to start mass production of typically 250,000 batteries annually by the mid–1990s.

8.2.6 Sodium–nickel chloride

Sodium–nickel chloride (Na/NiCl$_2$), or ZEBRA, and other sodium–metal chloride batteries are high–temperature devices that resemble the sodium–sulphur battery, the major difference being the positive electrode (which is an insoluble metal chloride in molten NaAlCl$_4$), and the temperature of operation (which is slightly lower). Benefits over the Na/S battery include the lower operating temperature, the ability to withstand limited overcharge and discharge, better safety characteristics, cell failures in short–circuit conditions and higher cell voltage. Disadvantages include a slightly lower energy density and lower power density (McLarnon and Cairns, 1989).

Work on the Zebra battery started only in 1975 in South Africa. The main development effort today is by Beta Research & Development in the UK, in collaboration with Daimler–Benz through its AEG subsidiary. The Mercedes–Benz 190 model passenger car has been used as a testbed for the Zebra battery.

8.2.7 Lithium aluminium–iron sulphide

Lithium aluminium–iron sulphide (LiAl/LiCl–KCl/FeS) cells operate at about 450°C, at which temperature the electrodes remain solid but the electrolyte is a molten salt. Compared with the sodium–sulphur battery, the LiAl–FeS battery displays a number of advantages, such as the ability to withstand numerous freeze–thaw cycles, cell failures in short–circuit conditions, the ability to withstand overcharge and low–cost construction techniques (ibid).

The LiAl–FeS system was originally developed in the early 1970s at the US Argonne National Laboratory. Today SAFT America, a subsidiary of the French battery company, is the leading developer of this technology and is being funded by the US Department of Energy (DOE) and the Electric Power Research Institute (EPRI).

8.2.8 Lithium aluminium–iron disulphide

The lithium aluminium–iron disulphide (LiAl/LiCl–LiBr–KBr/FeS$_2$) cell is closely related to the LiAl–FeS cell, and variations of it have been developed for a number of years. It uses a molten LiCl–LiBr–KBr salt electrolyte (melting point is 310°C) that permits operation at about 400°C, about 50°C lower than the LiAl–FeS cell. Early cell performance was comparable with that of sodium–sulphur, although recent cell tests have shown larger energy density and much larger power density, although how actual batteries perform is yet uncertain. The benefits listed in **8.2.7** for LiAl–FeS apply equally to LiAl–FeS$_2$ cells. Availability of a commercial LiAl–FeS$_2$ battery is expected to be at least 12 years away (McLarnon and Cairns, 1989).

8.2.9 Lithium–solid polymer

The rapidly growing demand for batteries having high specific energy and power has led to increased efforts in lithium battery technology. The inherent advantages of solid–state batteries in regard to safety and reliability are strong reasons for their continued development for EV applications. The ionic conductivity of the solid electrolytes, relative to liquid electrolytes, however, is low – resulting in lower power densities for solid–state systems operating at ambient temperatures. One advantage of lithium–solid polymer technology is that very thin cell construction is used (typically 44 microns total), leading to much greater flexibility in designing complete batteries to required shapes. Heated (to around 80°C) Li–solid polymer cells have performed well in regard to charging and discharging rates typically found in EV applications, and peak power densities have been as high as 1000 W/kg. The low cost of raw materials and low estimated cost of fabrication would appear to make these batteries economically viable for EV use, given further development (De Jonghe *et al*, 1991).

One of the leading efforts in lithium–polymer development is a joint venture formed between the Canadian Hydro Québec and Japan's Yuasa Battery Company, designed to develop the technology eventually to full–scale production. Lithium–polymer batteries for EV applications, however, are a long way from production and may only be considered as possible long–term options (EPRI, 1991). A Li–solid polymer battery has also been developed by AEA Technology at Harwell in the UK, and is currently being further developed with Dowty.

8.2.10 Zinc–air, aluminium–air and iron–air

Zn–air, Al–air and Fe–air batteries have all been considered as storage systems for EV applications, due to their high theoretical energy density. Low to moderate energy efficiencies are achieved, with the best values (about 60 percent) projected for the Zn–air system. Electrically rechargeable metal–air batteries rely on either a bifunctional air electrode (which must reduce O_2 on discharge and generate O_2 on charge), or a three–electrode system which has a cost, weight and volume penalty.

A fundamental problem with the favoured bifunctional system is the short lifetime of the electrodes (McLarnon and Cairns, 1989). The USDOE Battery Technology Research and Development Project (part of the Electric Vehicle Research and Development Program) is assisting with the development of metal–air batteries, focusing on improving the air electrode and scaling up the cell technology to an EV battery (USDOE, 1991).

8.2.11 Zinc–bromine

The zinc–bromine ($Zn/ZnBr_2/Br_2$) battery is similar to the zinc–chlorine battery (both zinc/halogen), the major difference being that the bromine evolved on charge is stored in a separate "oil" phase. Cells and batteries have often shown very short lifetimes, typically less than 200 cycles, often due to warping of electrodes. Additional research and development, particularly to identify chemically and dimensionally stable cell materials, is required. This system is still relatively novel and requires an elaborate pumping and heat management system. SEA of Austria has further developed the original Exxon system and its major advantage stems from its simplicity, cheap components and performance. Little EV testing has been conducted using this technology.

8.2.12 Nickel–metal hydride

Nickel–metal hydride cells are widely viewed as a superior replacement for commercial Ni–Cd batteries (DeLuca *et al*, 1991). Recent interest in this technology was heightened with the announcement that Energy Conversion Devices (ECD) in the US had been awarded a multi–million dollar contract from the US Advanced Battery Consortium to develop the technology, which Ovonic had been researching for 30 years previously. ECD claims the battery can be charged in as little as 15 minutes, can undergo more than 1,000 charge–discharge cycles and has a life of 10 years or 100,000 vehicle miles.

The fundamental difference between Ovonic's nickel–metal hydride battery and others is that one of its electrodes is made of nickel hydroxide and the other a disordered alloy of many metals – vanadium, titanium, zirconium, nickel, chromium and others, mixed during the metals' vapour phase. While solid electrodes react with the electrolyte only at their surface, the Ovonic electrode can react additionally throughout its interior because hydrogen ions are able to pass between the metal atoms, thereby improving considerably the charge retention that conventional cells store only at the electrode surface (New Scientist, 1992). Initial cell testing has demonstrated the potential for good performance – especially with regard to power density.

8.3   Electric vehicle technology

8.3.1   Electric and hybrid vehicle specifications

Table 19, commencing overleaf, and spanning pages 131 to 135, lists electric and hybrid vehicles that manufacturers have produced, either in running prototype or production form. The main vehicle specifications include, where known, the battery type and weight and the vehicle operating range and top speed. The vehicle range is usually based on a city/urban driving cycle, but where it is not apparent the quoted figures are presented. It is not an exhaustive list – new electric vehicle projects are reported frequently, and some recent ones, therefore, may be omitted.

Manufacturer	Vehicle description/ model	Passenger/ payload capacity (seats/kg)	Battery type/ weight (kg)	Battery life (discharge cycles/ years)	Recharge time (hours)	Range (km)	Top speed (km/h)
Audi AG, Germany	Car ✿ "Duo"	4–5	Ni–Cd 200 kg	–	Hybrid ¾	40 (battery only)	50 (batt only)
Auto Technik Walther GmbH, Germany	Car "Electro Microcar"	2	Pb–A 220 kg	–	8	60–100	75–90
BMW AG, Germany	Car ✿ "Electric 3 series"	4–5	Na–S	–	–	145	120
	Car ✿ "E1"	4	Na–S 200 kg	–	6–8	250	120
	Car ✿ "E2"	4–5	Na–S	–	6–8	430	121
California Electric Cars, Inc, USA	Car "Monterey"	2	Pb–A 470 kg	3–4 yrs	5–6	100–160	110
Carrozeria Autodromo Modena, Italy	Urban bus "Pollicino Etabeta"	20	Pb–A 1500 kg	1,200 2 yrs	8	50	55
Chrysler, USA	Minivan "Chrysler TEVan"	5	Ni–Fe 800 kg	–	8	>100	110
CityCom A/S, Denmark	3–wheeler "Mini–el City"	1½	Pb–A 90 kg	500 >2½ yrs	8–9	30–50	40
Clean Air Transit, USA	Shuttle bus "Electric Shuttle"	28	Pb–A 1862 kg	–	–	100	65
Clean Air Transport AB, Sweden	Car ✿ "LA301"	4–5	Pb–A or Na–S or Ni–Cd	3–4 yrs	Hybrid (8)	240 95 (Pb–A batt)	120
Conceptor Industries, Canada	Van "Electric G–Van"	5 680 kg	1270 kg	–	8–10	100	85
Coop Car/Microwet, Italy	Car "Smile Lady"	2	120 kg	4 yrs	6	35–45	80
Doran Motor Company, USA	3–wheeler ✿ "Doran Electric"	2	Pb–A 184 kg	200–400 1–3 yrs	5–7	70–100	135
Electricars Ltd, UK	Delivery float	1368– 2133 kg	Pb–A	–	–	30–80	35
Elroy Engineering Pty Ltd, Australia	Multi– purpose "Townobile"	9–15 600–900 kg	Pb–A 650–750 kg	>1250	4–8	>80	90
Fiat Auto SpA, Italy	Car "Cinque- cento Elettra"	2+2	Pb–A (or Ni–Cd)	600 (Pb–A)		70 (Pb–A) 100 (Ni–Cd)	80–85 (Ni–Cd)

Fiat Auto SpA, Italy	Car "Panda Elettra"	2	Pb–A 372 kg (or Ni–Cd)	700 3 yrs	8	60–70 (100)	70–80 (70–80)
	Car ✩ "Downtown"	3	Na–S	–	–	192–304	99
Ford Motor Company, USA	Van ✩ "Ecostar" (based on European Escort van)	2 410 kg	Na–S 363 kg	–	5½–6	>180 (400 – hybrid)	115
	Minivan "ETX–II"	7 500 kg	Na–S	–	8	160	105
Fridez–Solar AG, Switzerland	Car "Pinguin"	5	Pb–A 490 kg	>700	9	60–120	90
	Car "Pinguin"	2	Pb–A 350 kg	>700 >2 yrs	8	60–100	65–70
General Motors Corporation, USA	Car ✩ "Impact"	2	Pb–A 395 kg	20,000– 30,000 miles	2	190	160
	Van "GM G Van"	5 700 kg	Pb–A	–	8	60	85
	Van ✩ "HX3"	5	Pb–A	–	Hybrid (2)	150 (battery only)	130
	Car ✩ "XA–100"	5	Pb–A 298 kg	–	Hybrid (8)	570 – hybrid	105
General Motors – Opel AG, Germany	Car ✩ "Twin"	4–5	Li–carbon	–	–	–	120
	Car ✩ "Impuls 2"	4–5	Pb–A 395 kg	–	–	105	120
Genova Ricerche, Italy	Urban bus ✩ "L'Altrobus"	20	Pb–A 2000 kg	>3 yrs	None (hybrid)	–	55
	Minibus ✩ "Mini L'Altrobus"	8	Pb–A 650 kg	>3 yrs	None (hybrid)	–	60
HIL Electric Ltd, UK	Van "QT50"	2 500 kg	Pb–A/ Na–S	4 yrs	–	85–125	115
Hino Motors, Japan	City bus "HIMR"	–	Pb–A 563 kg Ni–Cd 380 kg	–	Hybrid	–	–
Horlacher AG, Switzerland	Car ✩ "City"	2	Pb–A 254 kg	–	5–8	50–120	80
	Car ✩ "Sport"	2	Pb–A 254 kg	–	3	50–150	120
Imatran Voima Oy/Neste Oy, Finland	Van "ELCAT Cityvan"	2 400 kg	Pb–A 440 kg	600 3 yrs	12	70–100	70
International Automotive Design, UK	Taxi ✩ "Eurotaxi"	6	Na–S	–	–	160	105

Iveco Fiat, Italy	Midi bus "Altrobus"	25 seats	Pb–A	–	Hybrid	–	55 (batt only)
	Minibus "Daily"	16–20 seats	Pb–A 960 kg Ni–Fe 652 kg	–	–	60 70	60 60
	Van "900E/E2"	430 kg	Pb–A 480 kg	–	–	60–80	60
	Van "Daily Electric E2"	1000 kg	Pb–A 980 kg	–	–	80	65
Kewet, Denmark	Car "El–Jet"	2	Pb–A	–	6–8	72–128	80
L&J Huntington Enterprises Pty Ltd, Australia	Car "Mira ECC"	2–4	360 kg	600 3–4 yrs	8–10	50–80	85
Larag, Switzerland	Car "Larel Elettrica"	2	Pb–A	–	–	–	80
Mazda Motor Corporation, Japan	Car ✿ "323 EV"	–	Pb–A	–	–	105	40
Mercedes–Benz AG, Germany	Car ✿ "190"	4–5	Na–NiCl$_2$ 330 kg	–	–	175	130
	Car ✿ "A–93"	4	Na–NiCl$_2$	–	–	160	–
	Van "MB180"	–	Zn–air 350 kg	–	–	200–300	–
Mitsubishi Motors, Japan	Car "Lancer"	4	Ni–Cd	–2000	8	200	110
	Car "Libero"	4	Pb–A Ni–Cd	–	8	103 155	128
	Car ✿ "ESR"	4	Ni–Cd	–	Hybrid	500	200
Nissan Motor Company, Japan	Car ✿ "Future Electric Vehicle"	4	Ni–Cd 200–250 kg	–	¼ fast 4 slow	160–250	130
Nordiska El–Fordon AB, Sweden	Van "Reliant Eco"	2–4 400 kg	Pb–A 360–480 kg	1000–2000 5–10 yrs	10	60–120	80
Piaggio VE, Italy	Van "Ape Elettrocar"	2 536 kg	360 kg	1000	8	–	44
PSA (Peugeot Citroën), France	Van "Citroën C25/Peugeot J5"	800 kg	Pb–A	–	5–6	75	90
	Car ✿ "Peugeot 205"	4	Ni–Cd 282 kg	–	8–10	115	100
	Car "Peugeot 405 Break"	4–5	Ni–Cd	–	Hybrid	72 (battery only)	130
	Van ✿ "Citroën C15"	–	Ni–Cd 352 kg	–	8–10	115	80
	Car ✿ "Citroën Citéla"	4	Ni–Cd	–	1–2	110	110

Renault, France	Van "Master Électrique"	750 kg 1000 kg	Pb–A Ni–Cd	–	–	80 120	80 80
	Car–derived van "Express Électrique"	360 kg 440 kg	Pb–A Ni–Cd	–	–	45 110	80 80
	Car "Electro–Clio"	4	Pb–A 420 kg	–	6–8	45	117
Rover Group, UK	Car ✿ "Metro"	4	Na–S 200 kg	–	–	112–136	–
Saab–Scania	Bus ✿	Single deck bus	Ni–Cd 680 kg	–	Hybrid	–	60
Sebring Auto–Cycle, Inc, USA	3–wheeler "Zzipper"	2	Pb–A 270 kg	2–5 yrs	6–8	85	90
Solar Car Corporation, USA	Car "Ford/Solar Festiva"	2	Pb–A 272 kg	1–5 yrs	6–10	40–95	105
Solectria Corporation, USA	Car "Solectria Force"	2 4	Pb–A 380 kg 294 kg	250–800 1–4 yrs	8 4–8	145–195 95–130	105 95
	Car ✿ "Solectria Flash"	2	Pb–A 199 kg	250–600 1–4 yrs	6–7	195–225	96
Solstar, Switzerland	Car "Solcar I"	2	Pb–A 230 kg	–	6–10	50	80
Spijkstaal Spijkkenisse BV, Netherlands	Van/Minibus "Spijkstaal/ VW Minibus"	9	Na–S	1000	5	140–200	105
Steyr–Daimler–Puch (Schweiz) AG, Switzerland	Car "Daimant"	2	Pb–A 256 kg	200	8	50	65
	Bus "Ely"	16	Pb–A 1300 kg	200	10	50	80
Torpedo SRL, Italy	Truck "Poker Elettrico"	2 550 kg	Pb–A 360 kg	1000 3 yrs	8	50–70	55
	Van "Terra Elettrica"	2 320 kg	Pb–A 360 kg	1000 3 yrs	8	60–100	60
	Car "Marbella Elettrica"	4	Pb–A 360 kg	1000 3 yrs	8	60–100	75
	Car ✿ "Torpedo Club"	4	Pb–A 420	1000 3 yrs	9	80–120	90
Toyota, Japan	Car ✿ "EV–50"	4	Pb–A	–	6–8	109–248	114
	Van ✿ "Town Ace EV"	4	Pb–A 670 kg	–	8	–140	85
Triple "O" Seven Corporation, USA	Car	2	317 kg	3–5 yrs	3–6	160	145
Volkswagen AG, Germany	Car ✿ "Jetta City– STROMer"	4–5	Na–S 276 kg	–	–	120	105

Volkswagen AG, Germany	Car ⚬ "Golf City-STROMer"	4–5	Pb-A 480 kg	–	–	56–81	–
	Car ⚬ "Golf hybrid"	4–5	Ni-Cd	–	Hybrid	20 (battery only)	60 (batt only)
	Car ⚬ "Chico"	4	Ni-Cd	–	Hybrid	320	130
Volvo Car Corporation, Sweden	Car ⚬ "ECC"	4–5	Ni-Cd	–	1.5 (hybrid) 6–15 (mains)	760 (hybrid) 85–146 (battery only)	175
W&E Electric Vehicles, UK	Delivery truck	1525–4065 kg	Pb-A	–	–	25–95	15–55

Notes:

Pb-A signifies lead-acid (or gelled lead-acid) battery
Na-NiCl₂ signifies sodium-nickel chloride (zebra) battery
Ni-Fe signifies nickel-iron battery
⚬ indicates running prototype/pre-production vehicle

Na-S signifies sodium-sulphur (beta) battery
Ni-Cd signifies nickel-cadmium battery
Zn-air signifies zinc-air battery

Table 19.   Electric and hybrid vehicle specifications

The data sources for Table 19 include Anderson (1990), Autocar & Motor (1991), Automotive Engineer (1991), Automotive Engineering (1992a, 1992b and 1992c), Barnett and Tataria (1991), Brusaglino (1991), CADDET (1991), Commercial Motor (1993 and 1994), Delarue (1992), Electric Vehicle Progress (1993a, 1993b, 1993c and 1994), Engineering (1993), EPRI (1991), Faust *et al* (1992), IAD (*undated*), Mechanical Engineering (1991), Mitsubishi (1993), PSA (*undated*), Reuyl (1992), Scott Keller and Whitehead (1992), SIP (1992), Suzuki *et al* (1992) and Volkswagen (*undated*).

8.3.2   Electric vehicle developments and demonstration projects

In the United States, the Department of Energy (USDOE) is undertaking a programme of near-term (<5 years), mid-term (5–10 years) and long-term (>10 years) research in transportation technologies, including batteries and fuel cells for EVs and hybrid vehicles (HVs) (Brogan and Venkateswaran, 1992). The programme aims to improve the performance, economics and reliability of EVs and HVs and make them competitive with conventional vehicles. A major driving force behind the growing research and development commitment to EV technology and the recent surge of industry interest in this area is the requirement for zero-emissions vehicles (ZEVs) by 1998 mandated by the State of California. These requirements are expected to be adopted by several other states.

Battery technology remains the critical barrier to the longer-term efforts targeted on full performance EVs with a 300–400 km range. Accordingly, the major portion of the USDOE programme funding is toward the $260 million research project with the US

Advanced Battery Consortium (USABC) to develop the most promising advanced battery technologies for EVs. This is a joint venture between the automotive industry, the electric utilities, battery manufacturers and the Federal government. A second major long–term thrust is provided by the USDOE Fuel Cell R&D programme, which is focused on the development of phosphoric acid technology for a hybrid bus application in the near–term, and proton–exchange–membrane (PEM) technology for light–duty vehicles in the mid to long–term.

In the US, various consortia have been established to pursue EV performance enhancement, to promote the conversion of defence industries to produce technologies for civilian EV applications and to create economic development opportunities. The consortia are supported by a $25M Defense Department program. The California Energy Commission is expanding its EV demonstration programme to include medium–duty and hybrid EVs. The program already supports the operation of several Conceptor G–Vans, the first certified EV to go into series production in the US. They are operated by local firms including the Pacific Gas and Electric Company (The Clean Fuels Report, 1991).

Each of the "Big 3" US automakers has expanded its existing EV programme in conjunction with the formation of the USABC. In addition to developing more advanced battery storage systems, the infrastructure developments are being researched, spearheaded in the US by the Electric Power Research Institute in California. While the effort is focused primarily on ensuring an adequate electricity supply for recharging EVs, issues such as standard voltages, currents and plug specifications are being established. A key concern to EV drivers – the time required for recharging the batteries, has received a considerable amount of research and development effort in the last few years.

The batteries of many EVs take between six and 10 hours to fully (re)charge, generally limiting the recharging period to overnight. Additionally, the prospect of exhausting the batteries before the end of a journey is of concern to potential EV users. The major developments required for charging systems, therefore, are in rapid recharging systems and inductive charging techniques. A rapid charging system developed by Nissan in conjunction with the "Future Electric Vehicle" is described by Fukino *et al* (1992). The *Super Quick Charging System* recharges nickel–cadmium batteries from a 400V, 140A supply to 40 percent of their capacity in six minutes. The batteries have been modified to reduce their internal resistance and a flat battery shape radiates the heat generated more quickly. A second rapid charging system has been developed for sealed lead–acid batteries, charging them to 40 percent capacity in 12 minutes. Other rapid charging systems are being developed by several manufacturers.

Inductive charging transfers electricity from a power source to a vehicle battery via a magnetic coupler. The potential benefits of inductive charging include universal charging, or the ability to charge any vehicle and battery regardless of manufacturer or battery type. Universal charging requires both a standard source inductor configuration and a smart vehicle-mounted charger that can meet the charging requirements of individual batteries. Because there is no direct transfer of current, inductive chargers minimise the risk of electric shock, and can be automated and made foolproof. Several inductive charging systems are under development, and are expected to be as efficient a conventional plug-in chargers (EPRI, 1992a).

In 1990 there were about 1,000 EVs in regular service in Japan (Iguchi, 1992) with new vehicles usually produced to order. However, a Ministry of International Trade and Industry (MITI)-sponsored EV programme has set a target of 200,000 EVs to be introduced into major cities by the year 2000 with objectives for range (250 km/charge), speed (120 km/h), battery life (4 years) and price relative to a conventional vehicle (no more than 20 percent higher). Battery development and government subsidy schemes have been initiated to encourage their purchase.

CITELEC is the European Association of Cities interested in Electric Vehicles, formed in February 1990, founded in the wake of the COST 302 subcommittee, where, from 1982 to 1986 eleven European countries studied the technical and economic conditions for the use of electric road vehicles (Maggetto *et al*, 1992). CITELEC intends to study the environmental, administrative and economic aspects of the introduction of EVs; to develop demonstration programmes for EVs; evaluate new vehicles on the market; to promote new development projects. A recent response to formation of CITELEC has been the IAD-designed "Eurotaxi" - a project to build a 21st century hybrid-electric taxi in collaboration with (so far) the cities of Oxford, Birmingham and Ipswich.

In the UK, City of Oxford Motor Services has won a contract to operate the first dedicated modern electric bus service in Britain. Operated under contract to Oxfordshire County Council, the route will involve four battery-powered Optare MetroRiders running between the railway station, the city centre and the science area of the University. Maximum charge is expected to provide the vehicles with a 50-mile range. The vehicles were converted by IAD of Worthing, in conjunction with Southern Electric.

Five European cities are to participate in a plan to rent EVs, on a short-term basis (by the hour or day). The Peugeot 106 and Citroën AX EVs will be based at city centre locations in Barcelona, Coventry, Karlsruhe, Milan and Tours. The programme is supported by the EC, the car company PSA and French public transport company Via GTI. The French city of La Rochelle has already commenced a large-scale EV demonstration project, with 25 Peugeot 106 EVs and 25 Citroën AX EVs. Recharging points have been installed around the city and are able to provide the nickel-cadmium batteries with sufficient charge in 10 minutes to travel over 12 miles (The Times, 1992).

A Swedish government-supported programme aims to start an EV market by seeking to get 1,000 EVs on the roads by 1997, with the first vehicles expected to be tested in 1995. The companies within the consortium include a car-leasing firm, taxi companies and the

Swedish Post Office. The cars and light vans will be used in three cities – Stockholm, Gothenburg and Malmo. In Finland, the city of Hämeenlinna is building 10 EV charging stations that will supply inexpensive electricity to as many as 15 Elcat Cityvans leased by local businesses. Another EV demonstration project was launched in 1992 on the Baltic island of Rügen in Eastern Germany with 14 cars in operation, expected to be extended to 60 by the end of 1993.

Four European companies are involved in an EC "Eureka" project to develop a city bus that runs on liquid hydrogen, used in fuel cells that are provided by the Belgian company Elenco. A nickel–cadmium battery, supplied by SAFT of France will act as a load levelling system and store the recovered braking energy. The project was due for completion during 1993 (The Engineer, 1991).

8.4   Electricity generation

EVs require electricity to power the motors that provide the vehicle propulsion. The electricity can be generated at the power station and distributed via the national grid to recharge batteries, or alternatively be generated on the vehicle. Longer–term, renewable energy may be partly provided by solar (or photovoltaic) cells. On–board electricity production may be from either a hybrid heat engine/generator/battery combination or from fuel cells which convert oxygen and hydrogen into electricity. This section deals with mains electricity and fuel cells – hybrid vehicles are discussed in the companion volume to this book, *Alternative Engines for Road Vehicles*.

8.4.1   UK power generation

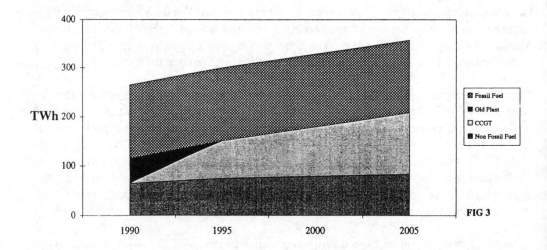

Figure 17.   Predicted UK electricity consumption and generation mix 1990–2005

Figure 17 (from Swinden and Johnston, 1992) shows the current UK electricity generation mix and demand and forecasts the change expected to 2005. As well as consumption predicted to increase by around 35 percent, the mix of power stations will change, based on current projects to build more natural gas–fuelled combined cycle gas turbine (CCGT) generators. The implications of future changes in the power station generating mix are discussed in section **8.5** – EV energy consumption and emissions.

8.4.2 Fuel cells

8.4.2.1 Introduction

Although fuel cells may be considered as alternative power sources, and could legitimately have been included in a review of alternative engines, it is considered most appropriate to include them in this fuels report, included in this section (electricity and EVs), since fuel cells provide electricity that EVs use.

A fuel cell is an energy conversion device, whereas a battery is an energy storage system. Because a fuel cell transforms the fuel (hydrogen is combined with oxygen) directly to electricity without combustion (and no therefore Carnot cycle efficiency limitation), there is little waste heat and a very high chemical–electrical energy conversion – typically 40 to 60 percent, based on fuel lower heating values (Swan and Appleby, 1992). The hydrogen used in automotive fuel cells can be produced from various sources, such as methanol or natural gas, which is converted into hydrogen and CO_2 through a thermal/chemical process in an on–board fuel reformer.

8.4.2.2 Fuel cell construction and types

A fuel cell consists of an electrolyte sandwiched between two electrodes; in a typical fuel cell, reactant gases are fed continuously to the negative electrode (anode) and positive electrode (cathode). All energy–producing oxidation reactions are fundamentally the same and involve the release of chemical energy through the transfer of electrons. *During combustion,* there is an immediate transfer of electrons, heat is released and water formed (since hydrogen and oxygen are the reactant gases). *In a fuel cell,* the hydrogen and oxygen do not immediately come together, but are separated by the electrolyte.

The electrons are first separated from the hydrogen molecule by a catalyst (reduction), thereby creating a hydrogen ion. The ion then passes through the electrolyte to the oxygen side, but the electrons cannot pass through the electrolyte and are forced to take an external electrical circuit (and provide useful work) which leads to the oxygen side. When the electrons reach the oxygen side, they combine with the hydrogen ion and oxygen to create water.

The fuel cell components (electrolyte, electrodes and catalyst), when assembled, are referred to as the electrode membrane assembly (EMA). In order to operate, the reactant gases at the appropriate pressure and humidity must be provided to each side of the EMA. The resulting electrons, water and heat from the reaction must be removed. Auxiliary systems are therefore necessary, including controls, cooling fans, recirculation pumps and often process air compression (since pressurisation significantly increases the electrode performance). The difference between gross fuel cell power and net power are the requirements of the auxiliaries.

The theoretical efficiency for the conversion of chemical energy into electrical energy in a hydrogen–oxygen fuel cell is 83 percent, although efficiencies of practical fuel cells using pure hydrogen and oxygen range from 50 to 65 percent (ibid). Fuel cells are usually classified by the type of electrolyte employed. These include the polymer electrolyte fuel cell (commonly known as proton exchange membrane fuel cell, or PEMFC), alkaline fuel cell (AFC), phosphoric acid fuel cell (PAFC), solid oxide fuel cell (SOFC) and molten carbonate fuel cell (MCFC). Some fuel cell types are suited to particular applications, such as combined heat and power (CHP) generation, or for larger power plants. The fuel cells most likely to find vehicle propulsion applications, and further described in the following sections, are the PAFC, PEMFC and SOFC.

8.4.2.2.1   Phosphoric acid fuel cell (PAFC)

PAFC is currently the only system which is available commercially and operates with an overall efficiency of 40 percent or higher. The system is based on immobilised phosphoric acid, operates at 180–200°C and uses air and hydrogen as fuel, thereby requiring external fuel reforming. Carbon and graphite are used extensively for construction, being virtually the only inexpensive materials not readily attacked by hot phosphoric acid. Platinum and platinum/base metal alloys supported on carbon are used as the catalyst. PAFCs have advantages over alkaline fuel cells in that they are tolerant of up to 2 percent (and perhaps even 5 percent) carbon dioxide and can be operated on reformed methanol, natural gas and naphtha. One major disadvantage of the PAFC is a slow start–up, in the region of 5 hours.

8.4.2.2.2   Proton exchange membrane fuel cell (PEMFC)

Of the various fuel cell types suitable for EV applications, the PEMFC system (or solid polymer fuel cell – SPFC) appears to be the strongest candidate (Swan and Appleby, 1992). The PEMFC is favoured because of its use of a solid electrolyte, cold start capability, relatively high power density and efficiency characteristics. The PEMFC was first designed by the General Electric Company for space flights as an auxiliary power source.

The PEMFC uses a solid polymer electrolyte that is manufactured in thin sheets, known as a membrane, comprising a polymeric backbone with side chains of sulphonic/sulphuric acid radicals. The fuel cell electrodes have a thin film of platinum catalyst supported on carbon which are bonded to the faces of the solid polymer electrolyte. Advantages of a solid polymer electrolyte include simplified sealing of the system during production, reduced corrosion and a longer cell and stack life. The PEMFC operated at below 100°C, and useful power can be drawn from it at room temperature, allowing a much faster and easier start–up than with the PAFC system (Automotive Engineering, 1992d).

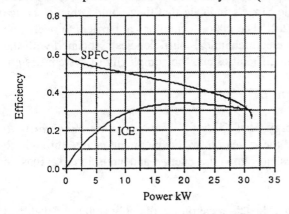

Figure 18 (from Adcock *et al*, 1991) alongside shows a comparison of the efficiency of a 30 kW PEMFC with methanol reformer, with a 30 kW internal combustion engine. The figure demonstrates the significantly higher efficiency of the fuel cell, especially at light– and part–load conditions.

Figure 18.   Efficiency of 30 kW internal combustion engine and PEMFC with reformer

In the UK, research is underway at Loughborough University on a solid polymer fuel cell–powered electric vehicle programme. Work includes the design of the fuel cell stack, the fuel management system, the vehicle layout and vehicle performance calculations. A comparison between a 40 kW power output, 1,000 cc internal combustion engine (ICE)–powered vehicle, a 250 kg sodium–sulphur battery–powered vehicle (30 kW) and a SPFC vehicle with methanol reformer (also 30 kW power output), as tested according to the EC urban cycle, has predicted a marginally lower acceleration and top speed for the EVs, although considered adequate for urban driving, and substantially better efficiency.

The ICE vehicle average efficiency was 16% whereas the fuel cell vehicle achieved 31% over the cycle. The weight of the ICE and transmission was 80 kg, the sodium sulphur battery was 250 kg (plus 130 kg for drive train) and the fuel cell components were 180 kg (plus 130 kg for drive train). While the battery–powered vehicle achieved the best overall efficiency (over 50%), this does not account for mains power generation and distribution efficiency, and at higher average speeds the efficiency will fall. Range for the ECE urban cycle simulation was predicted to be 770 km for the ICE vehicle, the SPFC vehicle was 730 km and the sodium–sulphur battery–powered vehicle was 190 km.

8.4.2.2.3   Solid oxide fuel cell (SOFC)

The SOFC is a high–temperature, compact ceramic device that operates at 1,000°C. Although much less developed than the PAFC and PEMFC, it has high fuel efficiency (in excess of 50%), high power density and uses low–cost materials in its construction. It can run on hydrogen and on alcohols or light hydrocarbons without an external reformer and does not require platinum or other exotic catalyst materials. The SOFC is expected to have rapid transient response to load variation and will not require a battery for peak power loads. Laboratory tests of single SOFCs indicate that they could achieve service lives in excess of 10 years (EPRI, 1992b). The major technological barriers to overcome to permit commercial viability involve electrochemical inefficiencies associated with the high operating temperatures, fabrication difficulties and the complexity of system design.

8.4.2.3   Future fuel cell development

The United States Department of Energy *Fuel Cells for Transportation Program* is to advance fuel cell technologies from the research & development stage through to demonstration in cars, vans and buses, to show the energy and emissions savings in relation to conventional fuels.

Near–term efforts are directed at phosphoric acid technology, the only suitably developed fuel cell for automotive applications at present (USDOE, 1992). The result will be a methanol–fuelled, fuel cell–powered bus with a performance equivalent to current diesel–powered buses, but with an expected reduction in emissions (at point–of–use) of more than 99 percent. The buses are expected to be delivered in 1993/94, evaluated during the course of 1994 and lead to prototype bus fleets from 1995 (Patil, 1992). An urban bus was selected for this program because its larger size can readily accommodate the packaging of a first–generation fuel cell–powered propulsion system, and the PAFC selected because of its near–mature state of development and because its operation on reformulated methanol has been demonstrated.

For the mid–term, the DOE Fuel Cells Program is directed at the introduction of proton exchange membrane fuel cells, since they can provide the power density required for cars and vans, given suitable further R&D to reduce costs and optimise performance. Research on the PEM fuel cell in recent years has improved the technology to a point where the performance necessary for automotive applications can be achieved with a hybrid PEM fuel cell/battery system. Vehicle demonstration with PEM technology alone is expected in the late 1990s. Ongoing fundamental research on materials and components for PEM fuel cells is directed at reducing their cost and improving their performance and endurance. Additional research is underway to improve on–board fuel reformers to permit methanol, ethanol, natural gas and other hydrocarbons to be converted to hydrogen as efficiently as possible.

The solid oxide fuel cell (SOFC) is expected to be targeted for long–term application. A feasibility study will evaluate design, performance characteristics, safety and cost factors.

One further area of USDOE Fuel Cell research is to investigate the direct electrochemical oxidation of methanol to permit reformerless fuel cell operation in vehicles. A direct methanol–air fuel cell (DMFC), capable of electrochemically oxidising methanol to CO_2 at the anode, is under consideration. This would provide benefits of weight and volume reduction, and would eliminate the need for cost–effective hydrogen storage technologies. These benefits would only be realised if a DMFC demonstrates performance characteristics comparable to those of the PEMFC. Strong worldwide interest has been demonstrated in the DMFC concept, and a targeted DOE program is likely.

8.4.3 Solar cells

The solar cell or photovoltaic (PV) cell is a semiconductor device which directly converts solar energy into electricity. The amount of current produced by a PV cell is proportional to the amount of light falling on it; thus current increases with the area of the cell and with the intensity of the light. The amount of solar energy reaching the earth in 20 minutes is equivalent to the amount of energy consumed on the planet in one year (Murthy, 1993).

Most commercial solar cells are crystalline silicon, for which efficiencies of 13–18 percent can be achieved, although present commercial thin film cells made from amorphous silicon may only demonstrate 5–6 percent efficiency, but are significantly cheaper to manufacture. New types of cell, such as cadmium telluride have shown efficiencies in excess of 12 percent. Single solar cells produce relatively small amounts of power, with a typical silicon cell of 100 cm^2 in area giving about 3A at 0.5V in full sunlight. Since this voltage level is insufficient for most applications, several cells are usually connected together to form a "module", typically containing 30–36 cells connected in series. Modules can also be connected to form an "array" if higher power is demanded.

Although scientists have known of the PV principle for many years, the technology has been prohibitively expensive until recently. Major developments in the last 15 years have lowered the cost of PV cells to the point where they can be economically preferable to conventional power supplies in some areas. From a solar electricity cost of US$2,000 per watt in the 1950s, it fell to about $60 per watt by 1976 and is currently at the $5 per watt level. Many experts are predicting that PV electricity is likely to cost $1–$2 per watt by the turn of the century (ibid). The main challenge for the future of PV electricity generation remains the reduction of cost (both by increasing solar cell efficiency and by reducing the cost of materials and construction), and most of the current research activities are related to this in some way. Considerable effort is being expended on system technology, including finding an efficient way of integrating a PV system with other electricity supplies, including the grid.

8.5   Electric vehicle energy consumption and emissions

8.5.1   Energy consumption

Comparing the emissions performance of EVs (or rather the emissions from the power stations that would usually generate the electricity used) with conventionally–fuelled vehicles is extremely difficult, given the present lack of experience of production–ready EVs with advanced batteries that are necessary to provide competitive range and performance. Most current EVs are substantially heavier than their petrol or diesel–fuelled counterparts, adding to increased energy usage per kilometre driven.

The implication for heavier vehicles is demonstrated in Figure 19 overleaf, which shows the relationship that exists between electrical energy consumption (measured in kWh/km) and the mass of the vehicle. The data has been gathered from a number of vehicle manufacturers and independent research organisations who have conducted tests over the SAE J227a "C" city cycle. The individual vehicles, and others, are listed in Table 20.

EV model (and battery type)	Vehicle mass kg	Energy consumption kWh/km		
		Constant 50/56 km/h	Constant 70/72 km/h	City cycle (SAE J227a C) or DIN 70030
Peugeot 205 car (Ni–Cd)	1,056	–	0.120	0.125
Rover Metro car (Na–S)	1,127	–	0.274 @ 72	–
Citroën C15 light van (Ni–Cd)	1,200	–	0.150	0.157
VW Golf CitySTROMer car (Pb–A)	1,410	0.200 @ 50	0.250	0.280 (DIN)
Chrysler/Pentastar TEVan (Ni–Fe)	2,634	0.153 @ 56	–	0.272
Vehma/Conceptor (GM) G van (Pb–A)	3,530	0.234 @ 56	–	0.335

Table 20.   Comparison of electric vehicle energy consumption figures

The performance of EVs, compared with conventional vehicles, is very dependent on driving conditions. For example, in slow–moving urban traffic, the internal combustion engine (especially petrol–fuelled) vehicle is very fuel inefficient, whereas in similar conditions the EV performs substantially better in terms of energy usage. Until enough data on the energy consumption of EVs in realistic operating conditions are available, energy and emissions comparison with conventional vehicles may be unreliable. Differences in the test cycles used also make inter–comparison difficult.

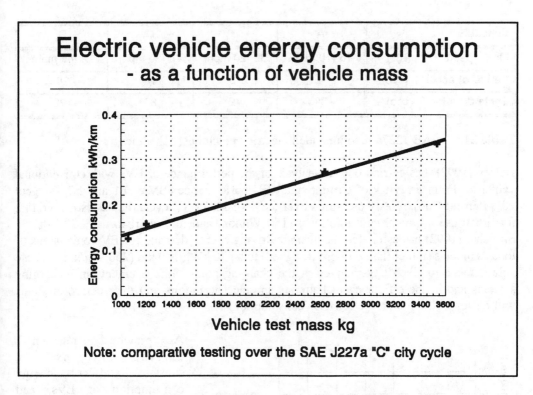

Figure 19. EV energy consumption as a function of vehicle mass

Many EV energy consumption figures are based on testing to the SAE J227a recommended test procedure (Amann, 1990), which involves acceleration from rest to a specified speed, a steady cruise, a deceleration to rest and a stationary period. Four cycles (or schedules) are specified in SAE J227a, depending on which best matches the vehicle's intended use. The maximum speed (V), time period in which to accelerate from rest to that speed (t_a) and the total test cycle duration (T), are shown in Table 21 overleaf.

From reviewing EV energy consumption data, it appears that the test most commonly employed is the "C" cycle, whose specified speed is 48 km/h (30 mph) and the time in which the vehicle accelerates from rest to this speed is 18 seconds. The "C" cycle, according to the SAE, is designed for testing a vehicle that is used over a variable route with medium frequency stop and go operation (for example, a parcel delivery van). The "D" cycle, more energy–demanding than "C", is intended for vehicles that are used over a variable route in stop and go driving typical of suburban areas. The adoption of the "C" cycle for determining EV energy consumption figures, with its relatively low speed and acceleration requirements, may explain why other test cycles reveal higher consumptions, such as VW indicate with energy consumption of the Golf CitySTROMer measured in accordance with DIN 70030 (see Table 20).

Schedule	A	B	C	D
Max. speed V (km/h)	16 (10 mph)	32 (20 mph)	48 (30 mph)	72 (45 mph)
Period of accel. t_a (sec)	4	19	18	28
Cycle duration T (sec)	39	72	80	122

Table 21. SAE J227a recommended electric vehicle test schedules

Delsey (1992), has carried out an analysis of the performance of EVs with conventional vehicles. From a review of current small EV performance (BMW E1 and E2, Peugeot 205, Renault Clio), mostly measured over the SAE J227a "C" cycle, he has estimated that for urban use a typical consumption is 150 Wh/km, and for suburban use 250 Wh/km. Streicher (1992) has estimated the average driving energy demand for EV conversions of today's models (rather than purpose designed) to be 26 kWh/100 km (260 Wh/km). These values relate to electricity metered at the charging point. Improvements in recharging systems and in the efficiency of batteries and electric motors and their control systems will all lower the metered electricity consumption.

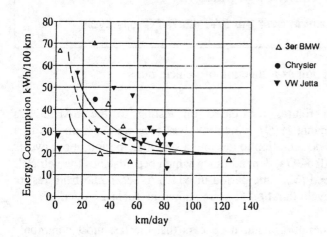

As previously mentioned, driving patterns can substantially affect the energy consumption of EVs, and Blümel (1992) illustrates how distance travelled can effect economy, as seen in Figure 20. The data was established during mixed city and highway driving of the vehicles shown on the key – they are generally medium sized cars.

Figure 20. EV energy consumption as a function of distance travelled

The EV energy values reported above are sharply contrasted by the energy demand of five small petrol–fuelled cars measured during the US FTP–75 drive cycle, designed to represent a mixed city/rural/motorway route. An average fuel consumption of 5.6l/100 km, equating to 50 kWh/100 km (500 Wh/km) energy was required as a result of the poor efficiency of the internal combustion engine (ICE) in typical use (ibid). On the face of it, therefore, it may appear that the EV is 2 to 4 times more energy efficient than the petrol vehicle, depending on its driving cycle. However, the efficiency of the electricity generation and distribution (via the grid) must also be accounted for, as should also the production and distribution energy for petrol and diesel fuel for conventional vehicles.

8.5.2 Emissions

It is the overall efficiency of the power station mix used and the feedstocks employed for the power generation that has caused several studies to conclude that the environmental benefits of EVs, while undisputed at point–of–use, are not conclusive in regional or global terms (Delsey, 1992, Streicher, 1992). Some studies (Blümel, 1992) conclude that EVs would substantially add to regional and global pollution, but all are highly sensitive to the particular electricity supply mix in the country considered and assumed EV usage pattern. Delsey and Streicher, for example, have considered the French and Austrian situation, while Blümel has considered the current German power generation mix.

Figure 17 shows the mix of UK power generation and how expected demand and the future mix of stations will change over the period 1990–2005. The notable difference is the increased use of natural gas as a fuel for combined cycle gas turbine generators. This, according to the electricity generating and supply industry (Swinden and Johnston, 1992), will have a positive environmental impact for the use of EVs recharged from a "cleaner" electricity supply.

The Energy Technology Support Unit has derived a set of greenhouse gas (GHG) emission factors for the generation of electricity, on the basis of a kWh delivered to the final user (Eyre and Michaelis, 1991, Martin and Michaelis, 1992). Emissions of CO_2, methane, NO_x, CO and non–methane HC (NMHC) are derived for electricity generation using coal (no emission control), best practice "clean" coal (with flue gas desulphurisation and low NO_x burners), best practice oil and natural gas–fuelled combined cycle gas turbine. Greenhouse gas emissions are derived for the various technologies based on both IPCC warming impacts and ETSU warming impacts (as a result of more recent modelling work at Harwell, it is thought that the warming effects of NO_x emissions was overestimated by IPCC).

Table 22 overleaf shows grams of CO_2 and total GHG emissions (in CO_2 equivalent) per kWh delivered, on a 50–year time basis, for a range of UK power generating technologies and includes both IPCC and ETSU warming factors. Also included are estimates of the CO_2 emissions from the proposed 1995 average and night mixes of UK power generation.

As previously stressed, comparing emissions from the use of EVs with those from conventionally–fuelled vehicles may not be reliable or more to the point, precise, given the little information of true "in–service" EV performance, which is expected to improve with advanced batteries and dedicated EVs anyway. However, a book such as this should provide some estimate of the potential impact from the use of EVs, specifically in the UK.

	IPCC warming impacts (50 yrs)		ETSU warming impacts (50 yrs)	
	CO_2	Total GHGs	CO_2	Total GHGs
Coal	938	1,410	938	990
Modern coal	976	1,340	976	1,050
Natural gas CCGT	431	510	431	430
1990 average mix	816	1,200	816	860
1990 night mix	738	1,110	738	780
Estimated 1995 average mix †	600	–	–	–
Estimated 1995 night mix †	500	–	–	–

† Based on Eastern Electricity predictions (Cooper Reade, 1992)

Table 22.   UK power generation greenhouse gas emissions: g/kWh delivered

Table 23 gives, for UK power generating technologies, the emissions caused by the use of a small electric car, as most data available is for this size of EV. Presented in g/vehicle km equivalent, the range of emissions is based on **an assumed EV power consumption of between 150 and 250 Wh/km.**

	Emissions: g / vehicle km equivalent			
	CO_2	THC	CO	NO_x
Coal	141–235	0.53–0.88	0.02–0.04	0.69–1.15
Modern coal	146–244	0.55–0.91	0.02–0.04	0.46–0.77
Natural gas CCGT	65–108	0.06–0.10	0.06–0.11	0.13–0.22
1990 night mix	111–185	0.42–0.70	0.02–0.03	0.54–0.90
Estimated 1995 night mix	75–125	–	–	–
Small 3–way catalyst-equipped car †	**132**	**0.09**	**0.87**	**0.19**

† from Streicher (1992)

Table 23.   Possible EV emissions performance for UK power supply mix

Table 23 indicates that EVs *may* provide an energy and emissions benefit in the UK, but this is highly dependent on several factors. Firstly, assumptions have been made of the energy consumption of EVs in use, and a range is suggested within which it is believed that *average* driving falls. However, even this suggested range of energy consumption may be inappropriate, dependant on the vehicle's actual usage and distance travelled. Figure 20 shows how measured energy consumption can vary with vehicle distance, and indicates that an EV used irregularly and for short distances may use quite high amounts of energy per kilometre driven. Studies of EV energy consumption in realistic operating conditions, not just test cycles (or at least if test cycles are used, they should include the same as for conventionally–fuelled cars to ensure inter–comparison) should be undertaken.

Another uncertainty is caused by not having representative vehicles for comparison. Current EVs are based on converted vehicles and are usually significantly heavier than their petrol or diesel–fuelled counterparts. Future purpose–designed EVs (as the conventional vehicle is purpose–designed for the ICE) with batteries at a more advanced stage than present, will be lighter and more efficient than those of today. Another factor to consider is how the future power supply will be generated. The planned increased use of natural gas, and emission factors for energy supplied from CCGT power stations, indicate that CO_2 and NMHC emissions may be lower, but NO_x and SO_x emissions could rise. In comparing the emissions shown in Table 23, it should be borne in mind that the 3–way catalyst–equipped petrol car's emissions are for a "hot" engine – or at least where the catalyst has reached light–off temperature. The issue of excess "cold–start" emissions, that would not be generated from the use of EVs, also requires investigation to fully understand the true environmental impact from the use of the EV.

8.6   Outlook

Electric vehicles have the advantage of emitting no emissions at point–of–use, and this has made them attractive in terms of addressing local air quality concerns. Air pollution in the Los Angeles basin has been the reason for the state of California mandating increasingly stringent emissions levels in the coming years, eventually implying the sole use of EVs.

Considerable progress has been made in EV battery and powertrain technology since the last surge of interest in the 1970s, but substantial development in EVs is still required to make them as attractive to motorists as conventional vehicles. If the development of advanced batteries continues as expected and their performance meets the criteria widely established as necessary for adequate EV range, acceleration and speed, a wider market for EVs *is* likely. The production cost (at present) of any EV is invariably higher, and usually much higher, than that of its conventional counterpart. If battery costs prove to be at the optimistic end of the ranges estimated, EV life–cycle costs may be lower than conventionally–fuelled vehicles. The California Air Resources Board (CARB) estimate that as EVs are mass–produced, the total cost increase per vehicle (relative to a vehicle of today, at current prices) would be about $1,350.

EVs in considerable numbers would reduce the demand for petroleum, the real cost of which is expected to increase substantially over the next few decades. EVs in considerable numbers would reduce air pollution, especially in towns and cities, and so would contribute to better human health and the conservation of buildings in urban areas. They are much quieter than conventional vehicles and so might be more acceptable in many situations, such as night time delivery duties. EVs are expected to have longer useful lives and require less maintenance than conventional vehicles.

Despite being a zero emission vehicle (ZEV) at point-of-use, the electricity used has to be generated somewhere. While renewable energy supplies should make no net greenhouse contribution, these represent a tiny fraction of the total generating capacity today, and for many years still to come. The advent of natural gas-fired power stations will reduce CO_2 emissions in future, and in the UK the future generating mix may prove attractive for EV usage. Depending on the energy requirements of EVs in practice, their use will confer local air quality benefits and may reduce national emissions of CO_2, although NO_x emissions could be higher.

8.7   Summary

The EV is a zero emission vehicle (ZEV) at point-of-use. They are much quieter than conventional vehicles and are expected to last longer and need less maintenance than conventional vehicles.

The key to their substantial introduction is the improvement in electricity storage, either in batteries or by using fuel cells to generate it on-board, and in reducing the cost of these technologies and the time needed to recharge them. EVs in considerable numbers would reduce the demand for fossil fuel, the real cost of which is expected to increase substantially over the next few decades. The global impact of EVs is variable, depending on the power sources and generating mix.

REFERENCES FOR CHAPTER 8

ADCOCK P L, A NEWBOLD, P J MITCHELL, P D NAYLOR and R T BARTON (1991). Solid polymer fuel cells for electric vehicle propulsion. Proceedings of the conference "Internal Combustion Engine Research in Universities, Polytechnics and Colleges", C433, Institution of Mechanical Engineers, London.

AMANN C A (1990). The passenger car and the greenhouse effect. SAE Technical Paper Series No 902099, Society of Automotive Engineers, Inc., Warrendale, Pennsylvania, United States of America.

ANDERSON W M (1990). An Electric Van with Extended Range. SAE Technical Paper Series No 900181, Society of Automotive Engineers, Inc., Warrendale, Pennsylvania, United States of America.

AUTOCAR & MOTOR (1991). Fiat on a charge. Pages 38–39, 7 August 1991.

AUTOMOTIVE ENGINEER (1991). Special feature: Electric traction. Pages 34–38, August/September 1991.

AUTOMOTIVE ENGINEERING (1992a). Global viewpoints: Concepts – PSA electric vehicles. Pages 55–56, Volume 100, No 6, June 1992.

AUTOMOTIVE ENGINEERING (1992b). Propulsion technology: an overview. Pages 29–33, Volume 100, No 7, July 1992.

AUTOMOTIVE ENGINEERING (1992c). Battery and electric vehicle update. Pages 17–25, September 1992.

AUTOMOTIVE ENGINEERING (1992d). Fuel cells: an overview. Pages 13–16, Volume 100, Number 4, April 1992.

BARNETT J H and H TATARIA (1991). Electric vehicle and battery testing at the Electric Vehicle Test Facility. SAE Technical Paper Series No 911917, Society of Automotive Engineers, Inc., Warrendale, Pennsylvania, United States of America.

BLÜMEL H (1992). CO_2 and pollutant emissions of catalyst–equipped, battery–powered and hybrid cars: A comparison. Proceedings of the International OECD/IEA Conference "The Urban Electric Vehicle: Policy options, technology trends, and market prospects", 25–27 May 1992, Stockholm, Sweden.

BRAITHWAITE J W and W L AUXER (1991). Status of the DOE–sponsored sodium/sulfur battery development program. Sandia National Laboratories, Albuquerque, New Mexico, United States of America.

BROGAN J J and S R VENKATESWARAN (1992). Diverse choices for electric and hybrid motor vehicles: Implications for national planners. Proceedings of the International OECD/IEA Conference "The Urban Electric Vehicle: Policy options, technology trends, and market prospects", 25–27 May 1992, Stockholm, Sweden.

BRUSAGLINO G (1991). Fiat development on electric vehicle technology. Seminar proceedings from "AutoTech 91", Institution of Mechanical Engineers, C427/20/117.

BURKE A F (1991). Battery availability for near–term (1998) electric vehicles. SAE Technical Paper Series No 911914, Society of Automotive Engineers, Inc., Warrendale, Pennsylvania, United States of America.

CENTRE FOR ANALYSIS AND DISSEMINATION OF DEMONSTRATED ENERGY TECHNOLOGIES (1991). Electric Vehicle Directory. Published in the USA by Electric Vehicle Progress, on behalf of CADDET, Sittard, The Netherlands.

CHEIKY M C, L G DANCZYK and M C WEHREY (1990). Rechargeable zinc–air batteries in electric vehicle applications. SAE Technical Paper Series No 901516, Society of Automotive Engineers, Inc., Warrendale, Pennsylvania, United States of America.

COMMERCIAL MOTOR (1993). Light vehicle news: Iveco gets friendly with hybrid Daily. Page 11, 7–13 January 1993.

COMMERCIAL MOTOR (1994). Vehicle news: Post Office charges. Page 8, 24 February–2 March 1994.

COOPER READE M (1992). Electric vehicles and the greenhouse effect. Paper from the seminar "Electric Vehicle Technology", 29 April 1992, Motor Industry Research Association, Nuneaton.

DE JONGHE L, S J VISCO and M M DOEFF (1991). Lithium/polymer batteries. Materials Sciences Division, Lawrence Berkeley Laboratory, Berkeley, California, United States of America.

DELARUE C (1992). The Renault pragmatic approach to cleaner European city cars. Proceedings of the International OECD/IEA Conference "The Urban Electric Vehicle: Policy options, technology trends, and market prospects", 25–27 May 1992, Stockholm, Sweden.

DELSEY J (1992). Environmental comparison of electric, hybrid and advanced heat engine vehicles. Proceedings of the International OECD/IEA Conference "The Urban Electric Vehicle: Policy options, technology trends, and market prospects", 25–27 May 1992, Stockholm, Sweden.

DELUCA W H, J E KULGA, R L HOGREFE, A F TUMMILLO and C E WEBSTER (1989). Laboratory evaluation of advanced battery technologies for electric vehicle applications. SAE Technical Paper Series No 890820, Society of Automotive Engineers, Inc., Warrendale, Pennsylvania, United States of America.

DELUCA W H, K R GILLIE, J E KULGA, J A SMAGA, A F TUMMILLO and C E WEBSTER (1991). Key battery test results at Argonne National Laboratory. Chemical Technology Division, Electrochemical Technology Program, Argonne National Laboratory, Argonne, Illinois, United States of America.

DELUCHI M A, Q WANG and D SPERLING (1989). Electric vehicles: Performance, life–cycle costs, emissions, and recharging requirements. *Trans Res – A*, Vol 23A, No 3, Pages 255–278.

ELECTRIC VEHICLE PROGRESS (1993a). Mitsubishi and Tokyo Electric have new EV on the streets. Page 2, 15 June 1993.

ELECTRIC VEHICLE PROGRESS (1993b). Subcompacts with EV option unveiled by Mercedes, BMW. Pages 1–2, 15 September 1993.

ELECTRIC VEHICLE PROGRESS (1993c). Mitsubishi ESR, and other EVs, shown at Tokyo Auto Show. Page 2, 1 December 1993.

ELECTRIC VEHICLE PROGRESS (1994). EV–50: This might be Toyota's California Car. Page 7, 15 January 1994.

ENGINEERING (1993). Electric Vehicles: Assault on batteries. Engineering, Vehicle Engineering & Design Supplement, Pages 7–10, May 1993.

EPRI (1991). Electric vehicles for the '90s. Electric Power Research Institute Journal, Pages 5–15, April/May 1991.

EPRI (1992a). Technical Brief: Inductive Charging. Electric Power Research Institute, Palo Alto, California, United States of America.

EPRI (1992b). Technical Brief: Solid Oxide Fuel Cell Research at EPRI. Electric Power Research Institute, Palo Alto, California, United States of America.

EYRE N J and L A MICHAELIS (1991). The impact of UK electricity, gas and oil use on global warming. ETSU Report AEA–EE–0211, Energy Technology Support Unit, Harwell.

FAUST K, A GOUBEAU and K SCHEUERER (1992). Introduction to the BMW E1. SAE Technical Paper Series No 920443, Society of Automotive Engineers, Inc., Warrendale, Pennsylvania, United States of America.

FUKINO M, N IRIE and H ITO (1992). Development of an Electric Concept Vehicle with a Super Quick Charging System. SAE Technical Paper Series No 920442, Society of Automotive Engineers, Inc., Warrendale, Pennsylvania, United States of America.

HENRIKSEN G L and J EMBREY (1991). Lithium metal sulphide technology status. Chemical Technology Division, Electrochemical Technology Program, Argonne National Laboratory, Argonne, Illinois, United States of America.

IGUCHI M (1992). Market expansion programme of electric vehicles planned by the Ministry of International Trade and Industry, Japan. Proceedings of the International OECD/IEA Conference "The Urban Electric Vehicle: Policy options, technology trends, and market prospects", 25–27 May 1992, Stockholm, Sweden.

INTERNATIONAL AUTOMOTIVE DESIGN (Undated). Eurotaxi. IAD, Worthing, Sussex.

MAGGETTO G, M LICCARDO and P VAN DEN BOSSCHE (1992). CITELEC. European Association of Cities interested in Electric Vehicles. Proceedings of the International OECD/IEA Conference "The Urban Electric Vehicle: Policy options, technology trends, and market prospects", 25–27 May 1992, Stockholm, Sweden.

MARTIN D J and L A MICHAELIS (1992). The environmental impact of electric vehicles. Proceedings of the International OECD/IEA Conference "The Urban Electric Vehicle: Policy options, technology trends, and market prospects", 25–27 May 1992, Stockholm, Sweden.

MCLARNON F R and E J CAIRNS (1989). Energy storage. Applied Science Division, Lawrence Berkeley Laboratory, Berkeley, California, United States of America.

MECHANICAL ENGINEERING (1991). Electric vehicles: Getting the lead out. Pages 36–41, December 1991.

MITSUBISHI MOTORS (1993). Present Status and Trend of Electric Vehicles. Mitsubishi Motors Technical Review 1993 No. 5. Mitsubishi Motors Corporation, Tokyo, Japan.

MURTHY M R L N (1993). Solar cells and their industrial applications. Semicon Tech Consultants, Bombay, India.

NEW SCIENTIST (1992). Technology: US battery research moves up a gear. Page 19, 30 May 1992.

PATIL P G (1992). Fuel cell for transportation – technology for 21st century. Proceedings from the International Conference "Next Generation Technologies for Efficient Energy End Uses and Fuel Switching", 7–9 April 1992, International Energy Agency/Bundesministerium für Forschung und Technologie, Dortmund, Germany.

PSA (Undated). PSA documentation: Electric vehicle and other areas of research. Peugeot Citroën, Paris, France.

REUYL J S (1992). XA–100 Hybrid Electric Vehicle. SAE Technical Paper Series No 920440, Society of Automotive Engineers, Inc., Warrendale, Pennsylvania, United States of America.

SCOTT KELLER A and G D WHITEHEAD (1992). Performance Testing of the Extended–Range (Hybrid) Electric G Van. SAE Technical Paper Series No 920439, Society of Automotive Engineers, Inc., Warrendale, Pennsylvania, United States of America.

SIP (1992). Press release: Ethanol and hybrid buses in extensive environmental test in Stockholm. SIP The Swedish International Press Bureau, Stockholm, Sweden.

STREICHER W (1992). Energy demand, emissions and waste management of EVs, hybrids and small conventional cars. Proceedings of the International OECD/IEA Conference "The Urban Electric Vehicle: Policy options, technology trends, and market prospects", 25–27 May 1992, Stockholm, Sweden.

SUZUKI T, T KOIKE, A OBATA and T TAJIMA (1992). Hino low emission and better fuel city bus with new diesel/electric hybrid engine. Hino Motors Ltd. and Toshiba Corporation, Japan.

SWAN D H and A J APPLEBY (1992). Fuel Cells and Other Long Range Technology Options for Electric Vehicles. Knowledge Gaps and Development Priorities. Proceedings of the International OECD/IEA Conference "The Urban Electric Vehicle: Policy options, technology trends, and market prospects", 25–27 May 1992, Stockholm, Sweden.

SWINDEN D J and P D JOHNSTON (1992). The potential for increasing energy efficiency in industry with a reduction in greenhouse gas emissions by switching to electricity. Proceedings from the International Conference "Next Generation Technologies for Efficient Energy End Uses and Fuel Switching", 7–9 April 1992, International Energy Agency/Bundesministerium für Forschung und Technologie, Dortmund, Germany.

THE CLEAN FUELS REPORT (1991). Electric Vehicles: Government Actions – CEC expands electric vehicle demo program. Pages 178–180, November 1991.

THE ENGINEER (1991). Techscan: Bus running on hydrogen timetabled for 1993. Page 36, 11 July 1991.

THE TIMES (1992). Pull up to the battery point. 24 January 1992.

US DEPARTMENT OF ENERGY (1991). Technology Factsheet No 9.0: Battery technology research and development. US DOE, Office of Transportation Technologies, Washington, DC, United States of America.

US DEPARTMENT OF ENERGY (1992). Electric and Hybrid Vehicles Program. 15th Annual Report to Congress for Fiscal Year 1991. DOE/CE–0357P. US DOE, Office of Transportation Technologies, Washington, DC, United States of America.

VOLKSWAGEN AG (Undated). VW Documentation: Forschung fur die Zukunft. CitySTROMer – Ein alternatives Antriebskonzept der Volkswagen–Forschung. Volkswagen AG, Wolfsburg, Germany.

WESTBROOK M H (1992). Electric vehicles – developments and limitations. Paper from the seminar "Electric Vehicle Technology", 29 April 1992, Motor Industry Research Association, Nuneaton.

9.   HYDROGEN

9.1   Introduction

Hydrogen is frequently dubbed the "ultimate fuel" since it generates no carbon dioxide during its combustion and is considered the cleanest transport fuel at point–of–use, barring electricity. Investigation into the use of hydrogen as a vehicle fuel began in the 1920s and 30s, when over a thousand vehicles were converted to hydrogen and hydrogen/petrol operation in Europe. For about three decades little attention was then paid to hydrogen but about 30 years ago the topic was revived and currently the strongest development efforts are centred in Japan and Germany (Holman *et al*, 1991).

Hydrogen can be stored for use in any of three ways. Liquid hydrogen (LH_2) is stored in cryogenic containers which tend to be bulky. Hydrogen can be compressed and stored in gas cylinders in much the same way as CNG, but at a much higher pressure – typically 70 MPa (700 bar or 10,000 psi). Alternatively, hydrogen can be bound with certain metals to form hydrides which are solid, stable compounds, which release the gas when heated. Another way in which hydrogen can power vehicles is via electricity provided by hydrogen–fuelled fuel cells in an electric vehicle. Fuel cells are discussed in chapter 8 (Electricity, section **8.4.2**).

9.2   Fuel characteristics

The thermal efficiency of a hydrogen engine should be at least 15 percent higher than its petrol–fuelled counterpart, mostly attributable to hydrogen's higher–than–petrol octane rating of 106 RON permitting the use of increased compression ratio (Milkins *et al*, 1987), although pre–ignition is more likely with hydrogen due to its very wide flammability limits. Power from a hydrogen–fuelled engine may be higher or lower than from a petrol counterpart; the major factor is the form in which the hydrogen is introduced into the cylinders. Gaseous fuel induction – external mixing, at ambient temperature displaces a significant amount of air, leading to a substantial power reduction of up to perhaps 50 percent, whereas direct injection of cryogenic hydrogen – internal mixing, will increase power output by 15–20 percent.

Operating very lean will increase efficiency (and reduce uncontrolled NO_x emissions) at the expense of power and driveability. Supercharging is one method of compensating for lost power in a lean–burn hydrogen engine. Hydrogen is much more susceptible to pre–ignition than petrol, necessitating the careful control of fuel and ignition timing. At high engine loads the combustion chamber temperature can rise significantly with hydrogen when supplied at ambient temperature. Water injection is usually necessary at full load to provide cooling and ensure smooth running and no pre–ignition (and can also reduce NO_x emissions by 50 percent) (Peschka, 1991).

Hydrogen's energy density is much lower than petrol and storage in compressed, liquid or hydride form requires very much bulkier and heavier tanks than for conventional fuels. On an energy–equivalent basis, the weight of hydrogen is 0.37 times that of petrol, while its volume is 3,000 times (Furuhama, 1992). At the stoichiometric air/fuel ratio, for example, hydrogen volume is 28 percent of the total mixture, whereas petrol is just 1.7 percent (Teramoto *et al*, 1992).

9.2.1 Safety implications

Despite potential safety problems, hydrogen is not considered a *particularly* dangerous fuel (US Congress, 1990). In the event of a fuel leak it will disperse (if not confined) very quickly in comparison with petrol which evaporates more slowly. Hydrogen is also non–toxic and is not carcinogenic. Stored in hydride form, major fuel leaks should not occur, adding to the safety of this form of fuel system.

Hydrogen flames are very hot, yet radiate little heat, making hydrogen fires difficult to locate. Hydrogen is an odourless and invisible gas, and although an odourant could be added to aid detection, this may contaminate hydride storage, if used. Contact with LH_2 destroys living tissue because of the very low temperature of $-253°C$, so serious cryogenic burning could arise from contact with hydrogen escaping from pressurised fuel systems.

A particular problem associated with the use of LH_2 is boil off. As the liquid warms, boil off gas is released which must be vented from the storage tank. Boil off in a confined space creates a high risk of fire or explosion since it is more likely to explode than an equal concentration of methane or petrol vapour if contacted by a flame (US Congress, 1990). This is due to its extremely wide flammability limits of 4 to 74 percent (compared with methane's flammability limits of 5 to 15 percent), making virtually any concentration of hydrogen in air liable to explode.

9.3   Feedstocks

Hydrogen can be manufactured from a wide source of feedstocks including water, coal, oil, natural gas and biomass. Hydrogen can be produced from natural gas in a steam reforming process which produces both hydrogen and carbon monoxide. Alternatively natural gas can be heated in the presence of a catalyst to be "cracked" into carbon and hydrogen. Coal or biomass can be gasified by combining it with steam under high pressure and temperature, forming carbon dioxide and hydrogen. Electrolysis of water or high–temperature steam electrolysis require large energy (electricity) inputs to generate hydrogen. A technique called photolysis uses light with a chlorophyll–type chemical to split water into oxygen and hydrogen.

Currently, steam reforming of natural gas is the cheapest production method (US Congress, 1990). Coal gasification may be the closest to becoming fully commercial for hydrogen manufacture although other systems, such as the Lurgi gasifier (for the production of methanol from natural gas), are fully commercial and produce synthesis gas – a combination of hydrogen and carbon monoxide.

It is widely agreed that the best long–term manufacturing process is the electrolysis of water. This process, at its current state of development, is very inefficient (far less efficient than recharging batteries, for example) and in order not to add to global pollution, the electricity for the process would need to be generated using renewable (or non–fossil) energy sources such as solar, hydroelectric or biomass. The use of nuclear power in the long term is presently under considerable debate.

9.4   Infrastructure

Theoretically it is possible to transport pure hydrogen "bullets" or mixtures of hydrogen and natural gas, with certain adjustments, in the existing natural gas pipeline network. In practice, however, this seems highly unlikely to be practicable, especially given hydrogen's propensity to leak through fractured or corroded pipes (Holman *et al*, 1991). It is more likely that a separate distribution network would be required using either pipelines or tankers. In either case this would add substantially to the cost of hydrogen distribution.

In the absence of a suitable pipeline infrastructure, the most efficient means of hydrogen distribution is in liquid form. The storage and safe handling of liquid hydrogen has reached an advanced technical standard, largely due to the developments in space programmes, especially in the USA. Large vacuum–insulated tanks exist in great numbers in the USA and LH_2 is transported in road and rail tankers and in cargo ships on a regular basis.

For refuelling purposes, fully–automated refuelling equipment has been developed and tested and is considered safe for non–experts to use. Refuelling of a 150–litre LH_2 tank (that is already cold) is possible in 3½ to 4½ minutes (Peschka, 1991). With hydride storage, refuelling to 90 percent maximum hydrogen capacity can be achieved in 10 minutes, 95 percent in 20 minutes and 100 percent about 45 minutes (Electronics World + Wireless World, 1991). It is usual, therefore, for hydride systems to be designed to be fully charged at 90 percent of their *actual* hydrogen storage capacity.

9.5   Vehicle modifications

Modified petrol and diesel (with ignition assistance) engines have been shown to be suitable for hydrogen operation. Early Daimler–Benz research with petrol engines used a dual–fuel system in which hydrogen was used at idle and an increasing fraction of petrol was separately injected at increasing load – to 100 percent petrol at full power (Dini,

1990). Although diesel dual–fuel operation is also possible, very little information is available as to its performance. Most hydrogen research, however, in both petrol and diesel engines, has been with a single fuel system.

To ensure satisfactory engine operation and to suppress blowback into the inlet manifold, multipoint injection is generally preferred, via mechanically, hydraulically or electromagnetically–operated injectors. Because of the larger volume of gaseous hydrogen required (than liquid fuel), fuel injection nozzles (or mixer unit if not injected) with multiple holes must be used. Specialised components must be designed to ensure the correct mixture preparation – whether the hydrogen is supplied at ambient temperature in gaseous form or injected as cryogenic hydrogen (high pressure direct cylinder injection or lower pressure port injection). If hydrogen fuelling is at ambient temperature using hydrogen stored as a liquid, an evaporation unit is needed, heated by the engine coolant.

The two methods for hydrogen storage that have received most research and development attention are as liquid hydrogen (LH_2) stored in cryogenic containers and bound with certain metals to form hydrides which are solid, stable compounds, which release the gas when heated.

Both systems have substantial limitations compared with petrol storage. Existing hydride storage systems must be very bulky because they can store only a few percent hydrogen by weight. For most materials, the weight of hydrogen stored is only 0.5 to 2 percent of the total weight of the storage tank, although a magnesium system will store as much as 3.6 percent by weight (US Congress, 1990). Ongoing research is aimed at developing a hydride storage system that can store a higher percentage of hydrogen by weight, and one developer has claimed seven percent using nickel–hydride in an amorphous form. This high a storage rate would make a hydride–based system much more competitive.

However, many alternative metal hydride systems being considered require a much higher pressure to charge them and a higher operating temperature that may be too high for the exhaust heat alone to provide (Electronics World + Wireless World, 1991). Peschka (1991) shows the mass of fuel required for equivalent energy storage to 25 kg (36 litres) of petrol, together with the required volume and complete storage system weight for both LH_2 and one form of hydride storage (iron–titanium hydride with a hydrogen capacity of 1.75% by weight) in Table 24 overleaf.

Advances in hydride storage technology would have a significant effect on the currently unacceptable tank weight (20 times that of petrol for iron–titanium) for equivalent vehicle range. The magnesium hydride system quoted earlier, with a hydrogen capacity twice that of iron–titanium (by weight), would halve the storage tank weight shown in Table 24. If a seven percent hydrogen capacity proves eventually to be possible, the tank weight would be around one–quarter that shown in the table at about 215 kg.

Fuel / storage system	Fuel mass / kg	Storage tank volume / litres	Storage tank weight / kg
Petrol	25	36.2	42
Liquid hydrogen (LH$_2$)	9.5	136	64.5
FeTiH$_2$ (hydride)	8.5	99	860

Table 24. Implications of hydrogen storage relative to petrol

A second possibility in hydride storage is that methylcyclohexane, a liquid hydride, might be employed as a hydrogen source. The hydride is stable at normal temperatures and pressures, so the only extra provisions to be made on the vehicle are a catalytic dehydrogenation unit, a pump to return the residue (toluene) and a tank to contain it after the hydrogen has been extracted. Although the system is too bulky and heavy for cars, it might be practicable for trucks and buses. Much research and development has been carried out by the Diesel Engine Research and Engineering Company (DERECO) in Switzerland. Since 1984 several trucks have been running on methylcyclohexane, although experimental dehydrogenation units have weighed 750 kg and 500 kg of fuel have been required for a 300 km vehicle range. Clearly, substantial further development is needed before this hydrogen fuel system reaches commercialisation (Automotive Engineer, 1991).

Cryogenic systems are not considerably heavier than petrol storage systems, so vehicle performance is not expected to suffer although its bulkiness will reduce vehicle space – even accounting for improved engine efficiency with hydrogen, a LH$_2$ storage tank would be several times the volume of a petrol tank for equivalent vehicle operating range. Further, the generally spherical or cylindrical shape of the tanks would be difficult to integrate into a vehicle design, although development of a square shaped tank is underway. Most LH$_2$ tanks suitable for development vehicles have a capacity of about 150 litres and have a mass of 50 to 60 kg. Constructed of aluminium alloy, they feature an evaporation–cooled vacuum super insulation system and suspension of the inner tank is via glass or carbon fibre–reinforced plastics. The blow–off, or leakage rate, is about 1.8 percent per day (Peschka, 1991).

9.6 Emissions performance

9.6.1 Vehicle exhaust emissions

As with all fuels, engine efficiency, performance and emissions from a hydrogen–fuelled engine are interdependent, and maximising one attribute may increase or decrease the others. Because hydrogen is simply that (it contains no carbon), no CO_2, CO or HC emissions are produced during its combustion, except from any lubricating oil or gas impurities that are present during combustion. The main combustion product is water which is emitted into the atmosphere as water vapour. Hydrogen vehicles would add very little to natural levels of evaporation, but whether small increases in ambient water vapour concentrations may have a disproportionately large effect on cloud cover and climate has yet to be assessed.

In general it should be possible to keep NO_x emissions at levels similar to, or below, those from a catalyst–equipped petrol vehicle using exhaust gas recirculation (EGR) with no exhaust gas aftertreatment, especially as hydrogen can be burnt in very lean mixtures. Cryogenic direct cylinder fuel injection offers the largest potential for achieving low-pollution combustion (Peschka, 1991), while retaining the operating characteristics of the engines with conventional fuels.

Welch and Wallace (1990) report that a direct injection diesel engine, converted to burn hydrogen, can emit lower levels of NO_x than a diesel–fuelled version, although injection timing is important and may possibly lead to equal or higher NO_x emissions if not optimised correctly.

9.6.2   Life cycle emissions

Life cycle emissions from the production (and use – although very small) of hydrogen have been estimated by DeLuchi *et al* (1988). Their analysis assumed the production of hydrogen using coal as the long–term feedstock and concluded that vehicles using hydrogen stored in hydride form would generate 100 percent *more* greenhouse gas (GHG) emissions than petrol–fuelled vehicles and that by using LH$_2$, the GHG emissions would be 143 percent higher. Coal gasification uses significant energy (which therefore generates CO_2 emissions) in the hydrogen–forming process, and *also* emits CO_2 as a product of the process itself, in an even larger quantity than that produced in supplying the process energy.

9.7   Costs

9.7.1   Fuel production and distribution

At present, a market for hydrogen in Europe hardly exists, as highlighted by the cost comparison between the USA and Europe, where present costs (in Germany) are some 3–5 times higher. European space projects, such as Ariane, are expected to reduce this considerably in the future. Costs for hydrogen produced at large US liquefying plants for space programmes, producing some 30 tonnes per day, are currently \$0.80 per gallon (Peschka, 1991).

9.7.2 Vehicle modification

The only vehicles running on hydrogen have been advanced prototype development vehicles. It is likely to be many years before significant numbers of vehicles are tested and even then the costs of modification will be extremely high. It is therefore very difficult to predict the cost of hydrogen–fuelled vehicles in comparison with conventional ones – although the costs, even of dedicated vehicles, are expected to be significantly higher due to the highly specialised components needed for the fuel storage and injection system.

9.8 Demonstration

Since the early 1970s, a number of LH_2 test vehicles have been developed and tested, mostly in the USA, Japan and Germany. In 1979 the first European LH_2 passenger car – a BMW 520, was demonstrated. Later BMW hydrogen test vehicles have included a 735i and 745i, developed in conjunction with Deutsche Forschungsanstalt für Luft und Raumfahrt (DFLR) of Stuttgart. In 1986 a BMW 754i was exhibited that was the first European vehicle with direct cylinder hydrogen injection. Special cryogenic injectors for the cold hydrogen (–233 to –253°C) and a LH_2 pump had to be developed for this application (Peschka, 1991).

Daimler–Benz have investigated hydrogen–fuelled vehicles employing hydride storage in fleet tests in Berlin since 1985 involving five passenger cars and five delivery vans. Due to a large storage volume necessary, the operating ranges of these vehicles have been low – typically 60–120 km. At the US Motor Show in Detroit in January 1991 the company produced a directly–fuelled hydrogen prototype car of the future – the F–100 (Cragg, 1992).

The Billings Energy Research Corporation in the USA carried out research into hydrogen–fuelled vehicles during the 1970s and converted a city bus, using hydrogen stored in iron–titanium hydride, for regular passenger service (Electronics World + Wireless World, 1991). In Canada, ORTECH and the University of Toronto have shown that a direct injection diesel engine, converted to burn hydrogen, can produce higher power output than a diesel–fuelled version (although overheating problems may occur) and emit lower levels of NO_X also (Welch and Wallace, 1990).

In Japan, the Musashi Institute of Technology has been involved in hydrogen–fuelled passenger car research since 1970, especially with direct cylinder hydrogen injection in 2–stroke petrol engines and diesel engines (2 and 4–stroke) with glow plug ignition assistance (Furuhama *et al*, 1987). Mazda Motor Company have demonstrated a hydrogen–fuelled rotary engine, claimed to be better than reciprocating engines for preventing backfiring and pre–ignition problems with hydrogen (Teramoto *et al*, 1992), and which does not suffer as badly from power reduction from the use of gaseous hydrogen at ambient temperature.

PSA (Peugeot–Citroën) in France investigated the use of hydrogen, stored in hydrides, as a vehicle fuel during the 1970s and culminated with extensive testing of a Peugeot 505 in the USA in 1981. Interest in hydrogen in the 1990s appears to be gathering momentum as PSA and Renault, in cooperation with the French Atomic Energy Commission (CEA), reassess its use with improved hydride technology (over that of the 1970s) and fuel injection systems (PSA, 1991).

In Australia, the University of Melbourne has converted a diesel engine to dual–fuel hydrogen operation, the ignition being started by a pilot injection of diesel fuel. More than 90 percent of the diesel is able to be substituted by hydrogen, and thermal efficiencies nearly 15 percent higher than the conventional engine were achieved. Power output was maintained or bettered (Lambe and Watson, 1993).

9.9   Outlook

The use of hydrogen as a vehicular fuel has strong appeal from a pollution control standpoint, and could assist in efforts to slow global warming if the hydrogen was produced from non–fossil fuel sources. However, much development work needs to be conducted before a hydrogen–based system could be practical, and the most likely cheapest system – a fossil–based one, would have a substantial negative greenhouse impact.

On the other hand, if photovoltaic–based electricity generation becomes successful, solar–based hydrogen could eventually be cost–competitive with coal–based hydrogen. Significant developments are being made in solar electricity production – Texas Instruments claims to be getting closer to Southern California Edison's domestic tariff rate of 14 cents/kWh with current solar power valued at 30 cents (Cragg, 1992). However, it is expected to take until at least the second decade of the 21st century before solar power can compete on cost with other energy sources. One other long–term possibility, if nuclear fission reactors are eventually developed, is to use their waste heat for thermal dissociation of water very economically (Automotive Engineer, 1991).

Apart from cost, hydrogen's major drawback may prove to be its storage bulkiness – hydrogen's low energy density implies either very limited range between refuelling or very large, heavy fuel tanks. Unless there is a major breakthrough in hydride storage or in vehicle efficiency, hydrogen–fuelled vehicles cannot provide a close substitute for petrol–fuelled ones.

An alternative use for hydrogen as a vehicular fuel is in fuel cells that generate electricity to power the vehicle via electric motors. The conversion efficiency of hydrogen into electricity can be 50 to 60 percent (Seiffert and Walzer, 1991), and because modern electric motors can demonstrate high efficiencies, the fuel cell–electric motor vehicle propulsion system would display a higher overall vehicle efficiency than today's drivetrains that suffer from the inherent Carnot cycle inefficiencies of the internal combustion engine (Langley, 1983). This use of hydrogen is discussed in chapter 8 – Electricity, section **8.4.2**. Given the need for important scientific and technological development in hydrogen manufacture and storage systems, however, hydrogen must be considered as a very long–term prospect as an alternative motor fuel.

9.10 Summary

Hydrogen is a gas possessing a high octane rating (RON=106). It can be used as both a petrol and diesel fuel substitute. It is non–toxic and disperses rapidly in the open.

Advantages of hydrogen

Hydrogen's high octane should permit higher thermal efficiency from a dedicated engine. Lean fuelling will increase the efficiency. No carbon compounds are formed from the combustion of hydrogen – just from lubricating oil or gas impurities. CO and HC emissions are very low. NO_x levels depend on the fuelling – lean–burn will produce similar levels to those from a catalysed petrol engine, if EGR is employed, and is the most efficient use of the fuel. Cryogenic direct injection of hydrogen offers the potential of very low emission levels.

Disadvantages of hydrogen

Gaseous fuel induction leads to a significant power loss (up to 50%). Storage of hydrogen can be achieved through compression, liquefaction or adsorption, but all these imply increased weight, volume and cost compared with conventional fuel tanks. More complex, specialised fuel components are required.

REFERENCES FOR CHAPTER 9

AUTOMOTIVE ENGINEER (1991). Special feature: Prospects for gaseous alternative fuels improved for LPG and LNG. Pages 38–41, February/March 1991.

CRAGG C (1992). Cleaning up motor car pollution. New fuels and technology. Financial Times Management Report, London.

DELUCHI M A, R A JOHNSTON and D SPERLING (1988). Transportation fuels and the greenhouse effect. Transportation Research Record 1175, National Research Council, Washington, DC, United States of America.

DINI D (1990). Energy/environment potential impact of hydrogen fuelled engines operating on road vehicles. Proceedings from the XXIII FISITA Congress "The Promise of New Technology in the Automotive Industry", 7–11 May 1990, Turin, Italy.

ELECTRONICS WORLD + WIRELESS WORLD (1991). Science: The hydrogen economy. Pages 668–671, August 1991.

FURUHAMA S (1992). Trend of social requirements and technological development of hydrogen–fuelled automobiles. *JSAE Rev*, Vol 13, No 1, January 1992.

FURUHAMA S, M TAKIGUCHI, T SUZUKI and M TSUJITA (1987). Development of a hydrogen powered medium duty truck. Proceedings of the 4th International Pacific Conference on Automotive Engineering "Mobility: The technical challenge", 8–14 November 1987, Melbourne, Australia.

HOLMAN C, M FERGUSSON and C MITCHELL (1991). Road transport and air pollution: Future prospects. Rees Jeffreys Discussion Paper 25, Transport Studies Unit, Oxford University.

LAMBE S M and H C WATSON (1993). Optimising the design of a hydrogen engine with pilot diesel fuel ignition. *Int. J. of Vehicle Design*, Vol. 14, No. 4, pages 370–389.

LANGLEY K F (1983). The future role of hydrogen in the UK energy economy. ETSU Report R15, Energy Technology Support Unit, Harwell.

MILKINS E E, H C WATSON, Z H ZHOU and J EDSELL (1987). Comparison of ultimate fuels – hydrogen and methane. Proceedings of the 4th International Pacific Conference on Automotive Engineering "Mobility: The technical challenge", 8–14 November 1987, Melbourne, Australia.

PESCHKA W (1991). Hydrogen as a fuel for earthbound vehicles. Proceedings of the conference "Engine and Environment – Which fuel for the future?", 23–24 July 1991, Grazer Congress, Graz, Austria.

PSA (1991). A Plan for the Environment – Other Areas of Research. Peugeot Citroën Communications Department, Paris, France.

SEIFFERT U and P WALZER (1991). Automobile technology of the future. Society of Automotive Engineers, Inc., Warrendale, Pennsylvania, United States of America.

TERAMOTO T, Y TAKAMORI and K MORIMOTO (1992). Hydrogen fuelled rotary engine. Proceedings from the XXIV FISITA Congress, 7–11 June 1992, Institution of Mechanical Engineers, London.

US CONGRESS (1990). Replacing gasoline: Alternative fuels for light–duty vehicles. Office of Technology Assessment, US Government Printing Office, Washington, DC, United States of America.

WELCH A B and J S WALLACE (1990). Performance characteristics of a hydrogen–fuelled diesel engine with ignition assist. SAE Technical Paper Series No 902070, Society of Automotive Engineers, Inc., Warrendale, Pennsylvania, United States of America.

10. COMPARATIVE EVALUATION OF ALTERNATIVE FUELS

10.1 Introduction

This chapter makes a comparative overall evaluation of alternative fuels with conventional crude oil–derived petrol (and diesel) in a number of areas. The point–of–use CO_2 emissions are compared and estimates of the ranges of fuel–cycle greenhouse gas emissions are shown. The fuel–cycle analysis takes account of all the energy and emissions associated with the fuel production, distribution, combustion and subsequent environmental impact. An indication of the effects of alternative fuels on the regulated (and certain unregulated) emissions are shown. The implications to the vehicle fuel storage system is discussed with the volume and weight of alternative fuel tanks compared. A comparison of the estimates of costs of alternative fuels is made, but this information is essentially outside the scope of the research that contributed to this book and so is not covered in great detail.

A discussion of the environmental and health effects of various pollutants (those from petrol, diesel and alternative fuels) is presented as a complete and self–contained chapter in this book – chapter 11.

10.2 Energy–specific CO_2 emissions (at point–of–use)

Table 25 overleaf shows, for a range of conventional fossil–derived fuels and alternatives (regardless of their derivation), the energy–specific CO_2 emissions at point–of–use, that is those exhausted from the engine after combustion. The table gives the chemical "formula" or composition of each fuel together with its energy content and subsequent CO_2 produced during combustion, on a mass basis. Energy–specific carbon dioxide emissions are shown in the fifth column of Table 25. This is the amount of CO_2 produced per unit of energy (or heat) that the fuel gives up during its combustion. The position of the alternatives are compared with petrol (taken as 100%) in this analysis.

Certain alternative fuels are able to be used more efficiently in the internal combustion engine that petrol, due to a range of factors (such as higher thermal efficiency by increasing the compression ratio to take account of a fuel's higher octane rating, improved volumetric efficiency and decreased compression work due to a fuel's higher heat of vaporisation, or by improved engine warm–up efficiency). This means that the engine can produce the same work (or power output) from a smaller energy input (less fuel). However, specific engine modifications (or dedicated engine designs) are required to realise these advantages fully. If these changes can be engineered, the relative engine operating efficiencies of certain alternative fuels are estimated (from Ho and Renner, 1990, IEA, 1990 and Martin and Shock, 1989) as: **100%**: Petrol; **110%**: LPG (Propane) and CNG (Methane); **115%**: Methanol, Ethanol (and Diesel).

Fuel	Chemical "formula"	Energy content[1] MJ/kg	CO_2 kg/kg fuel	CO_2 kg/MJ	CO_2 /MJ w.r.t. petrol
Coal	70% $CH_{0.8}$	28	2.88	0.103	140%
Crude oil	$CH_{1.8}$	42	3.19	0.076	103%
Diesel	$CH_{1.75}$	43	3.20	0.075	101%
Petroleum	$CH_{1.85}$	43	3.18	0.074	100%
Ethanol	C_2H_5OH	27	1.91	0.071	96%
Methanol	CH_3OH	20	1.38	0.069	93%
Propane (LPG)	C_3H_8	46	3.00	0.065	88%
Methane (natural gas)	CH_4	51	2.75	0.054	73%
Hydrogen	H_2	121	0.00	0.000	0%

[1] Based on lower heating value (LHV). The higher heating value is used for calculating process energies where water from combustion is condensed and the latent heat recovered. Vehicles do not condense (much) water in their exhaust and no heat is recovered; hence the LHV is used.

Table 25.   Specific heat output and carbon dioxide emissions at point of use for various fuels (Source: Ho and Renner, 1990, Kempe's, 1992 and Waters, 1992)

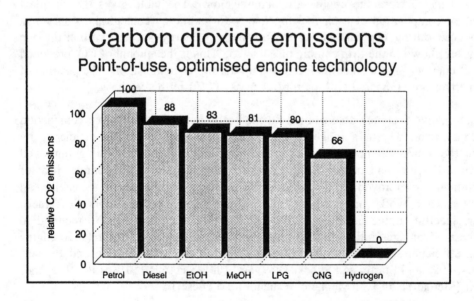

Figure 21.   Point–of–use CO_2 emissions from the combustion of alternative fuels for optimised engine designs

The relative engine work–specific CO_2 emissions (that is taking account of possible engine efficiency improvements) are shown in Figure 21. Engineering changes to conventional petrol engines can also bring about increased operating efficiency in the future, but since many of the changes are able to be applied to alternatively–fuelled engines also, the relative efficiency of the alternative fuels quoted is unlikely to change much – these improvements are a function of the properties of the fuels themselves.

Figure 21 shows that for point–of–use CO_2 emissions for alternative fuels used in optimised engines, diesel produces about 88 percent of those from light–duty petrol engines, ethanol 83 percent, methanol 81 percent and LPG about 80 percent. Of the various alternative carbon fuels, methane (natural gas) displays the lowest levels of CO_2 emissions at about two–thirds (66 percent) those from petrol–fuelled engines. It should be noted that these values are, strictly speaking, theoretical *engine* emissions, and that different vehicle types and different driving cycles will cause fluctuations in actual vehicular CO_2 emissions. Hydrogen produces no CO_2 during combustion.

10.3 Fuel cycle CO_2 and greenhouse gas emissions

10.3.1 Greenhouse gas emissions and their effects

With reference to the impact alternative fuels have on the "greenhouse effect", it is necessary to consider more than just point–of–use (i.e. combustion–related) CO_2 emissions, and to take account of all energy inputs to (and subsequent CO_2 emissions), and fuel losses from, the fuel cycle, by including the fuel production and refining processes and distribution of the fuel to the filling stations. Furthermore, CO_2 is not the only gas to contribute to the "greenhouse effect".

Two other gases which have a direct and significant effect are methane (CH_4) and nitrous oxide (N_2O) which are also emitted into the atmosphere during the production, distribution and combustion of fuels. Other direct greenhouse gases, not necessarily emitted as a result of fuel production or usage, include water vapour, ozone (O_3) and chlorofluorocarbons (CFCs). Gases which have a direct greenhouse effect may also have an indirect effect, be it positive or negative.

Certain other gases have an indirect effect, such as carbon monoxide (CO) which destroys hydroxyl radicals (OH) in the atmosphere, preventing them from scavenging methane (which has a radiative forcing effect many times higher than that of CO_2). Oxides of nitrogen (NO_x) and non–methane hydrocarbons (NMHCs) also play an indirect role, although the role of NO_x is under debate, and is thought may even have a negative indirect effect (Michaelis, 1991).

The overall greenhouse impact of a gas depends on both its concentration in the atmosphere and its effectiveness at absorbing heat (known as its radiative forcing effect).

The life (or period of residency) of a gas also has an effect on the overall greenhouse impact. The contribution a gas makes to the "greenhouse effect" is generally assessed over a specific time period, such as 20, 50 or 100 years, by integrating the effect that the release of a quantity of a greenhouse gas has over that time period.

The Intergovernmental Panel on Climate Change (IPCC, 1990 and 1992) has proposed a set of Global Warming Potentials (GWPs) for greenhouse gases, to serve as a measure of the possible warming effect on the earth surface–troposphere (lower atmosphere) system arising from the emission of each gas relative to CO_2. Table 26 shows the IPCC GWPs over a 100–year time span with the revised potentials issued in 1992 as a result of better understanding of estimated lifetimes of certain gases.

Gas	Global Warming Potential (100 year horizon)		Indirect effect
	1990 estimate	**1992 estimate**	
Principal anthropogenic greenhouse gases			
Carbon dioxide (CO_2)	1	1	none
Methane (CH_4)	21*	11*	positive
Nitrous oxide (N_2O)	290	270	uncertain
CFC – 11	3,500	3,400	negative
CFC – 12	7,300	7,100	negative
HCFC – 22	1,500	1,600	negative
HFC – 134a	–	1,200	none
Other (indirect) anthropogenic greenhouse gases			
Carbon monoxide (CO)	3	2.5†	positive
Nitrogen oxides (NO_x)	40	0†	uncertain
Non–methane hydro-carbons (NMHC)	11	5†	positive

* Including indirect positive warming effect
† ETSU (Michaelis, 1991) provisional revised estimate

Table 26. IPCC estimates of Global Warming Potentials for various gases

The IPCC has also shown the estimates of the relative contribution (both direct and indirect) of anthropogenic greenhouse gas emissions, integrated over a 100–year time horizon, using the GWPs. It shows how gases with an indirect effect (such as CO, NO_x, NMHCs) play their role in contributing to enhanced global warming. This is shown in Figure 22 overleaf.

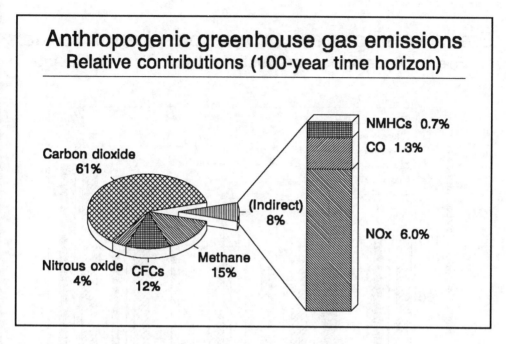

Figure 22. The relative greenhouse contributions from all man–made emissions (from IPCC, 1990)

10.3.2 Alternative fuels – fuel cycle greenhouse gas emissions

Several studies of alternative fuel cycles have been carried out, and their conclusions are shown graphically in Figure 23 overleaf. All of the studies compare greenhouse gas (GHG) emissions with those of petrol, and the studies have mostly compared the effects from passenger cars. As Figure 23 notes, absolute comparison between the different GHG emissions from the various alternative fuels is not reliable due to the different assumptions made of GWPs, process and distribution efficiencies, fuel losses and vehicle test criteria.

The fuel cycle analysis carried out by DeLuchi *et al* (1988a) compares petrol with natural gas and with methanol (using natural gas and coal as feedstocks) for fuelling *all* highway vehicles (including heavy–duty) in the United States. It shows that using compressed natural gas (CNG) as a vehicle fuel would have an overall GHG reduction of 19 percent relative to petrol. The use of methanol produced from natural gas would have a marginal GHG reduction of three percent, but if coal was used as the methanol feedstock, the GHG emissions would rise by between 52 and 98 percent, depending on the conversion process efficiency (based on a 30 percent more efficient process than current, which is defined as the baseline).

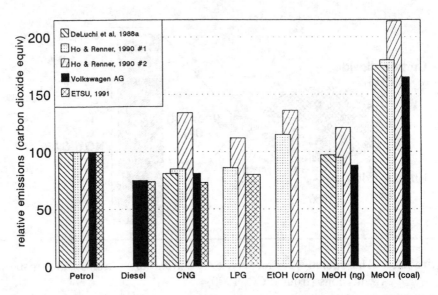

Figure 23.   Estimates of ranges of greenhouse gas emissions (CO_2 equivalent) from the fuel cycle analysis of alternative fuels

The study (ibid) estimated emissions of GHGs from the vehicles and from the production and distribution of the fuels themselves (DeLuchi *et al*, 1988b). The GWPs assumed for CH_4 and N_2O were 11.6 and 175 respectively (over a 75–year time period). The higher efficiency achieved by the use of dedicated alternatively–fuelled engines have been accounted for (13 percent for both methanol and natural gas), as has the extra (or reduced in the case of natural gas) energy used in the distribution of the fuel itself. The extra weight of gas storage cylinders in the vehicles has also been accounted for.

Some of the assumptions and estimates made by DeLuchi *et al* (1988a and 1988b) were on the basis of very limited data (such as certain vehicle emissions). Natural gas pipeline losses were put at three percent of the total delivered. The analysis is moderately sensitive to the GWP of methane – with the range of 5 to 30 being quoted as plausible, the overall GHG advantage of CNG would range from 25 percent to four percent better than petrol as a vehicle fuel.

Ho and Renner (1990) from the Amoco Oil Company have also estimated the greenhouse impact of using alternative fuels in passenger cars by carrying out a fuel cycle analysis. Their GHG emissions relative to petrol are shown in Figure 23 for two distinct scenarios that were considered. These two cases, a base case and an advanced case, are detailed below:

- Base case   1990 base year petrol–fuelled vehicle (for comparison) has assumed fuel economy of 29.0 mp(US)g. Alternatively–fuelled vehicles (with CNG, LPG, methanol and ethanol) are dual–fuel without optimisation – fuel consumption adjustment made for relative fuel heating values only. GWP for CH_4 = 50, N_2O = 300. Natural gas transmission losses = 1.5% (CNG) and 0.75% (for methanol); and

- Advanced case   2010 petrol–fuelled vehicle has assumed fuel economy of 35.0 mp(US)g (20.7% improvement). Alternatively–fuelled vehicles have dedicated engines with higher thermal efficiency than petrol (10% higher for CNG and LPG and 15% higher for the alcohols). GWP for CH_4 = 10, N_2O = 200. Natural gas transmission losses = 0.5% (CNG) and 0.25% (methanol). Process conversion efficiencies are also improved so as to emit between 35 and 55 percent less GHGs in CO_2 g/mile equivalent.

This alternative fuel cycle analysis (ibid) shows that using CNG as a vehicle fuel ranges from 34 percent worse to 15 percent better than petrol in terms of GHG emissions. While the base scenario is reasonable in the assumption that dedicated alternatively–fuelled engines do not exist, and that higher thermal efficiencies cannot be realised, the GWPs assumed for methane and nitrous oxide are probably too high. The revised estimates from the IPCC (1992) for the GWP for CH_4 and N_2O are 11 and 270 respectively. Thus the warming potential of the methane released into the atmosphere from transmission losses and as unburned fuel may be over–estimated by a factor of five.

The two estimates from Ho and Renner of the GHG emissions from the use of LPG are 12 percent worse to 14 percent better than petrol, based on the same two scenarios. However, their analysis assumed natural gas as the feedstock for the LPG, rather than crude oil, and LPG being a subsequent refinery product. Their analysis, therefore, assumed natural gas transmission losses which would enhance the global warming impact from this analysis. For LPG produced from crude oil the GHG emissions would be lower.

The estimates of GHG emissions from the use of methanol produced from natural gas are 21 percent worse (base case) and five percent better (advanced case) than from petrol. Where methanol uses coal as a feedstock, the GHG emissions range from 114 to 80 percent worse (even allowing for a 43% reduction in emissions due to process efficiency improvements) than for petrol. The study (ibid) also considered ethanol produced from corn, which demonstrated a GHG emissions disadvantage with respect to petrol, ranging from 36 to 15 percent, mostly due to the fuel production and processing.

Volkswagen AG (*undated*) has presented data relating to fuel cycle GHG emissions for alternative fuels, and is also shown in Figure 23. Again, petrol is taken as the baseline fuel and various alternatives compared with it, presumably for passenger cars (or other light–duty vehicles) since the equivalent emissions are presented in g/mile based on the US FTP–75 driving cycle. Despite not knowing all the assumptions made for the analysis, such as the GWPs assigned to the various greenhouse gases, the overall impacts are in line with most other studies. CNG has an advantage of 19 percent, methanol from natural gas an advantage of 12 percent, while methanol produced from coal shows an increase of 65 percent in GHG emissions. Also included in the VW analysis is diesel fuel (from crude oil) which demonstrates a 25 percent GHG emissions advantage over petrol.

The GHG fuel cycle emissions from the last study shown in Figure 23 (ETSU, 1991) for diesel, CNG and LPG, relative to petrol (used in a passenger with three–way catalytic converter), are also in line with most of the previous studies. Diesel shows a 26 percent advantage, CNG a 27 percent advantage and LPG a 20 percent advantage. While the ETSU study included GHG emissions from the production and supply of fuels, it is unclear how much (if any) allowance was made for transmission losses – especially that of natural gas.

Although Figure 23 shows that some considerable variations exist between the studies, a general trend is apparent. If the base case (base scenario) in the Ho & Renner study is not included (partly for reasons discussed above), the general estimates of the greenhouse impact of alternative fuels are found to be very similar. By taking the use of petrol as the reference, the use of diesel shows an advantage of about 25 percent, CNG an advantage of around 20 percent, LPG an advantage of some 17 percent and methanol (produced from natural gas) an advantage of about six percent. Where coal is used as a feedstock for methanol, an additional 75 percent GHG emissions are produced.

The reported case of the 15 to 36 percent worse greenhouse impact of using a biofuel (ethanol produced from corn) cannot be taken as a reliable estimate for all biofuels – the true greenhouse impact depends on how the crop is grown, how much and which fertiliser is used, how much energy (and whether fossil fuel or biofuel–derived) is used in the crop harvesting and processing and distribution of the fuel in addition to the vehicular emissions themselves. Certain claims that biofuels have a zero overall greenhouse impact are probably misleading. While it is true that the carbon released to the atmosphere during combustion can be absorbed by replanting the same quantity of crops used to produce the fuel, emissions produced from other crop growing and fuel production and distribution aspects can be significant. Studies that show biofuels as having no net greenhouse impact, therefore, are probably only referring to the carbon cycle as a consequence of growing and then burning the crop (fuel).

10.3.3 Alternative fuels – emissions from production and distribution as a proportion of total greenhouse gases

Several of the analyses of fuel cycle emissions described in section **10.3.2** have broken down the GHG emissions into the contributions from fuel production and processing and distribution, in addition to the vehicle emissions during combustion. Figure 24 shows the percentage of total fuel cycle GHG emissions that the *fuel production and distribution* contributes. The remainder is from combustion.

Greenhouse gas emissions from alternative fuels
Fuel production and distribution contribution

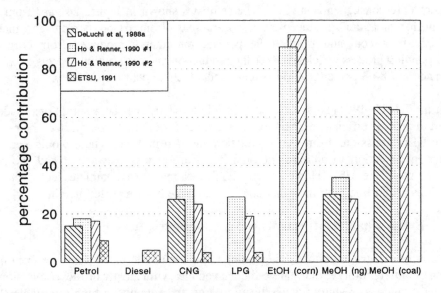

GHG emissions from fuel production & distribution shown as % of total

Figure 24. Alternative fuels – fuel production and distribution contribution to total greenhouse gas emissions

The estimates of the emission of GHGs due to the production and distribution of petrol range from nine to 18 percent of the total GHG emissions due to the use of this fuel. The estimates from the US–based studies show this contribution to be around the 15–18 percent mark, while the ETSU (1991) study puts this at nine percent. Both DeLuchi *et al* (1988b) and Ho and Renner (1990) have estimated the production, processing, transport and distribution process efficiency of petrol (defined as the net energy delivered per energy content of the fuel) as between 83 and 86 percent. Why, therefore, the ETSU estimate is so much lower would require further investigation. The only estimate for diesel fuel is from the ETSU study which shows five percent of the total GHG emissions arising from the production and distribution.

For CNG the estimated percentage of GHG emissions arising from production and distribution range from three to 32 - the US-based studies at 24-32 percent (assuming process efficiencies of between 76 and 81 percent) and ETSU estimating three percent. The percentage for LPG ranges from four to 27 percent - again the US studies show a much higher value of between 19 and 27 percent (process efficiency of between 79 and 84 percent) and the ETSU study four percent.

The process conversion efficiencies for alcohols are much lower than for other alternative fuels which means the contribution from the fuel production and distribution to total GHG emissions are consequently higher. The estimates shown in Figure 24 vary from 26-35 percent for methanol produced from natural gas, from 61-64 percent for methanol produced from coal and from 90-94 percent for ethanol produced from corn. The corresponding process efficiencies for the production and distribution aspects of these are estimated at 58-63 percent, 50-64 percent and 21-29 percent respectively.

From the ETSU (1991) study it is not possible to determine the assumptions made with regard to process efficiencies, fuel losses and other factors that would affect the estimates of the GHG emissions from fuel production and distribution. There would need to be further investigation to address the significant differences between it and the studies carried out in the US. However, the ETSU estimates of complete fuel cycle GHG emissions are in line with the other studies that have been reported herein.

10.4   Exhaust (pollutant) emissions from alternative fuels

Much of the exhaust emissions data relating to the use of alternative fuels are from limited testing of initial prototype alternative-fuelled vehicles, with mostly bi-fuel conversions (not optimised engines) resulting generally in a large and sometimes highly variable (higher or lower) range of emissions values. Consequently, much of the relative emissions performance is best expressed qualitatively, as shown in Table 27 overleaf (from Chang *et al*, 1991, IEA, 1990, Singh *et al*, 1987 and TRRL, 1992). The Table gives a qualitative summary of the regulated (and some currently unregulated) vehicle exhaust pollutant impacts of using alternative fuels, with reference to petrol.

The baseline fuel for comparison, petrol, is assumed to be used in an engine with a three-way catalytic converter. Studies of emissions from alternative fuels have almost always been made relative to this baseline. Certain fuels, such as rapeseed oil methyl ester (RME), are used as a diesel fuel substitute and other fuels, such as methanol and natural gas, can be used in both spark and compression-ignition (usually petrol and diesel-fuelled) engines. Therefore certain emissions performance comparisons should be made relative to diesel fuel and not simply petrol, as Table 27 just shows.

Fuel	Exhaust emissions performance					
	CO	NO$_x$	THC	NMHC	Form– aldehyde	Acet– aldehyde
Petrol (3 way–cat)	baseline	baseline	baseline	baseline	baseline	baseline
Diesel	lower	higher	= / lower	lower		
Reformulated petrol	lower	=	lower	lower		
Methanol	=	±	lower	lower	higher	=
M85	= / lower	= / lower	lower	lower	higher	=
Ethanol	=	±	lower	lower	=	higher
Natural gas	lower	±	= / higher	lower		
LPG	lower	±	lower	lower		
Hydrogen	lower	= / lower	lower	lower	none	none

THC: total hydrocarbons   **NMHC:** non–methane hydrocarbons   **=:** approx. equal emissions
±:   variable (depending on fuelling strategy; lean–burn or stoichiometric)

Table 27.   Pollutant emissions performance of alternative fuels

In general, all the alternative fuels listed in Table 27 result in lower vehicle exhaust emissions of non–methane hydrocarbons (NMHCs) and lower total hydrocarbons (THCs), with the exception of natural gas which results in higher methane emissions. The effect on CO is more varied, with either equal or lower exhaust emissions resulting – the alcohols (or petrol/alcohol blends) may result in little CO reduction relative to petrol. The air/fuel mixture has a crucial role to play – a leaner (than stoichiometric) mixture will reduce CO, but prevent the effective use of a three–way catalytic converter. NO$_x$ emissions can be lower, the same or higher, dependent on the fuel.

Alcohols can have lower NO$_x$ emissions due to their internal cooling effect, but in a dedicated engine with increased compression ratio the NO$_x$ emissions can rise to similar to those from petrol. Natural gas, LPG and hydrogen all tend to increase the exhaust emissions of NO$_x$ if higher compression ratios are used to take advantage of the higher octane ratings. However, using these fuels at a stoichiometric air/fuel ratio with a properly developed three–way catalyst may reduce the NO$_x$ emissions to a level approaching those from three–way catalysed petrol engine. Low NO$_x$ emissions (but probably still higher than from a three–way catalyst–equipped petrol) can be obtained by employing lean air/fuel mixtures with these fuels.

While diesel engine emissions of CO, HC and NO_x are lower than that from an uncontrolled petrol engine, the use of a three-way catalytic converter lowers the petrol NO_x emissions to below those from diesel fuelled-engines. Particulate emissions from diesel are many times those from petrol. Substitutes for diesel include vegetable oils (for example rapeseed oil methyl ester that may be used as a direct substitute), natural gas and alcohols (usually methanol). The main advantage is the lower emission of particulates; the combustion of methanol, ethanol and natural gas produce virtually none. CO emissions can be variable - generally higher with alternatives unless they are burnt in a lean mixture (which then precludes the use of a three-way catalytic converter). NO_x emissions can be variable - depending on fuelling (for example whether dedicated fuel or bi-fuel) and whether used in an optimised engine. Generally, though, diesel engines can be designed so that NO_x emissions decrease with increasing amounts of alcohol.

More aldehydes are produced as a consequence of burning alcohols (rather than petrol or diesel). Formaldehyde from methanol and acetaldehyde from ethanol are a cause for possible concern, especially as the aldehyde fraction can be about 10 percent of the unburned alcohol, compared with one percent of petrol (IEA, 1990). The health and environmental effects of aldehydes is discussed in chapter 11 of this book.

One notable consequence of using many alternative fuels, whether or not they appear to exhibit significantly lower exhaust emissions, is the reduced reactivity of many of the HC (or NMHC) emissions, and lower emissions of volatile organic compounds (VOCs). A large proportion of the hydrocarbon figure for alcohols represents oxygenates, which are less reactive in the atmosphere than the long-chain hydrocarbons they replace. These emissions can result in lower secondary pollutant formation, such as that of ozone and other photochemical oxidants. However, because the use of some alternative fuels can increase NO_x emissions, the exact interaction of different amounts of NO_x and HCs and their effect on photochemical reactions and the generation of secondary pollutants is not fully understood.

10.5 Vehicle storage implications of alternative fuels

The implications for on-board storage of alternative fuels have been discussed in each relevant chapter. Nevertheless, a brief comparative summary is provided in the form of Figure 25 overleaf (from Volkswagen AG, *undated*) which shows the volume required to store each alternative fuel, plus the weight of the storage system, for an equivalent vehicle operating range provided by that from 55 litres of petrol - typical of that stored in a medium-sized European passenger car.

The weight of the fuel storage system for petrol and diesel fuel are about identical at 46 kg, with the diesel requiring a lower volume for equivalent vehicle range - complete storage system volume of 60 litres versus 67. Vegetable oil (RME) would need about the same space requirement as petrol but would cause an overall weight increase in the region

Vehicle on-board storage of alternative fuels
Fuel tank weight and volume comparison

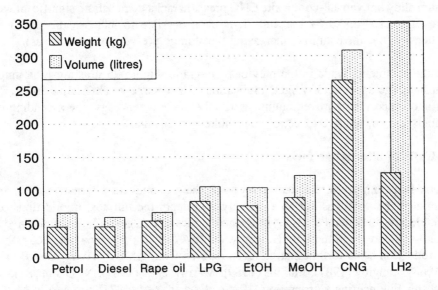

Note: based on equivalent range to that provided by 55 litres of petrol

Figure 25.   Vehicle on–board storage implications of alternative fuels

of 17 percent.   For LPG the weight and volume requirements increase by 80 and 57 percent respectively to 83 kg and 105 litres.

Ethanol would require about the same space as for LPG but with a lower overall storage system weight of 76 kg (65 percent above that for petrol). Methanol, with an energy content lower than ethanol, would therefore require more storage space (121 litres) and weight more (88 kg) than the petrol vehicle.

To use compressed natural gas and achieve identical vehicle range as petrol, the storage system weight and space required are 263 kg and 308 litres – increases of 470 and 360 percent respectively. The assumption is made that the CNG storage pressure is 165 bar, but this can be, and frequently is, higher (such as 20 MPa or 200 bar), with a consequent reduction in space. The use of hydrogen as a vehicle fuel, stored in liquid form in a cryogenic tank, would require 350 litres of storage space and weigh 124 kg.

It is not clear what assumptions VW have made with regard to the derivation of these figures. It is possible, for example, that the space required and weight of many alternative fuel storage systems will fall in the future as a result of improved engine efficiencies (as with dedicated designs) and lightweight storage tanks. For example, fibre–reinforced aluminium alloy or even all–composite CNG pressure tanks demonstrate significant weight saving over steel – up to 57 percent. It is even possible to increase the stored fuel's energy density (by, for example, increasing the storage pressure of natural gas).

Despite the comments made in the previous paragraph, Figure 25 illustrates the range of on–board storage space and weight implications for a range of alternative fuels. Even making allowance for improved engine and fuel storage technology, the ascending order of the fuels shown, relative to petrol, are unlikely to change.

10.6   Costs of alternative fuels

This book is essentially a technical analysis of the potential for alternative fuels and is based on a review that was not able to analyse in depth the full cost implications of the use of alternatives. Several analyses of the economics of using alternative fuels have been made, although mostly in the US. However, the IEA (1990) presents a range of estimates of the production costs for alternative fuels, derived from AMEC (1987), CNEB (1986), DOE (1988), Langley (1987), Markus (1989), Marrow *et al* (1987), NEDO (1987), Stone and Webster Engineering Corporation (1985), Wagner *et al* (1987) and World Methanol Conference (1987). The IEA study was therefore chosen to be highlighted in this book since the input from a number of European (and other) countries are included.

The fuels covered in the IEA report, relative to petroleum from crude oil, are CNG, very heavy oil (VHO) and tar sands products (such as petrol and diesel fuel), methanol (from natural gas, coal and biomass), ethanol from biomass and synthetic petrol and diesel fuel (both from natural gas). While the IEA data are presented in 1987 US$ on the basis of a barrel (bbl) of petroleum energy equivalent, for this report the baseline (taken by the IEA as crude oil at $18 per bbl used to produce conventional petrol at $27/bbl) is assigned a value of 100 and all other costs are relative. Since 1987 the price of crude oil has shown large fluctuations but appears to have stabilised – the current price of North Sea Brent crude is around $19/bbl (The Times, 13 November 1992), only five percent higher than in 1987, the year in which the IEA report is economically–based.

The IEA report shows costs for production and also overall costs, that is including distribution of the fuel. These two cost elements have been separated in order to show how certain fuels, such as alcohols and natural gas, would require substantial infrastructure investment, while very heavy oil (VHO) products and petrol or distillate synthesised from gas do not. The addition of incremental distribution and end–use costs does not alter the overall economic ranking order, but does bring synthetic petrol into closer competition with methanol. The data is considered sufficiently representative of IEA member

Alternative fuel overall costs
Basis: barrel of petroleum energy equivalent

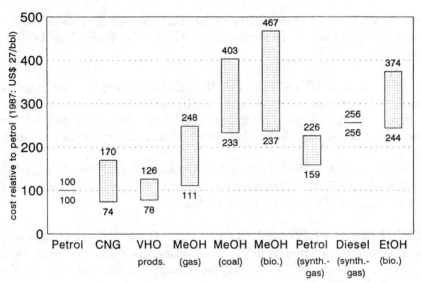

Figure 26. Estimates of the ranges of overall alternative fuel costs

countries' work in the field, and the ranges are themselves sufficiently broad that plants built today would be expected to produce these fuels at costs falling within these ranges.

Figure 26 shows the ranges of estimates for the overall costs of the alternative fuels quoted by the IEA. The conclusions from the study are that except for the lower end of the cost range for CNG and VHOs, all the alternative fuels are more expensive (in 1987 terms) than petrol. The situation, therefore, based on 1987 economics, is that only CNG and VHOs may be competitive given the 1987 oil price (which was roughly 1½ times that of today's *[August 1994]* price of US$17-18/bbl) and 1987 fuel technologies. However, feedstock-fuel conversion processes for many alternative fuels have improved considerably in recent years (IEA, 1990), with the natural gas-methanol process estimated to achieve 30 percent or even 50 percent cost reduction in the long term. In the years since 1987, therefore, significant process efficiencies and cost reductions may have been implemented which may alter the range of costs shown in Figure 26. The cost aspects of producing and distributing alternative fuels needs to be re-evaluated using current economic and technological data.

REFERENCES FOR CHAPTER 10

AUSTRALIAN MINERALS AND ENERGY COUNCIL (1987). Report of the Working Group on Alternative Fuels. AMEC.

CANADIAN NATIONAL ENERGY BOARD (1986). Canadian energy supply and demand 1985–2005. CNEB, Canada.

CHANG T Y, R H HAMMERLE, S M JAPAR and I T SALMEEN (1991). Alternative transportation fuels and air quality. *Environ Sci Technol*, pp 1190–1197, Vol 25, No 7, 1991.

DELUCHI M A, R A JOHNSTON and D SPERLING (1988a). Methanol vs. natural gas vehicles: A comparison of resource supply, performance, emissions, fuel storage, safety, costs, and transitions. SAE Technical Paper Series No 881656, Society of Automotive Engineers, Inc., Warrendale, Pennsylvania, United States of America.

DELUCHI M A, R A JOHNSTON and D SPERLING (1988b). Transportation fuels and the Greenhouse Effect. Transportation Research Record 1175, National Research Council, Washington, DC, United States of America.

ENERGY TECHNOLOGY SUPPORT UNIT (1991). Research and technology strategy to help overcome environmental problems in relation to transport. Final report, study group 2, under contract to DG XII of the European Commission. ETSU, Harwell Laboratory, Oxfordshire.

HO S P and T A RENNER (1990). Global warming impacts of gasoline vs. alternative transportation fuels. SAE Technical Paper Series No 901489, Society of Automotive Engineers, Inc., Warrendale, Pennsylvania, United States of America.

INTERGOVERNMENTAL PANEL ON CLIMATE CHANGE (1990). Climate change. The IPCC scientific assessment. World Meteorological Organisation/United Nations Environment Programme, Cambridge University Press, Cambridge.

INTERGOVERNMENTAL PANEL ON CLIMATE CHANGE (1992). 1992 IPCC Supplement. World Meteorological Organisation/United Nations Environment Programme, Cambridge University Press, Cambridge.

INTERNATIONAL ENERGY AGENCY (1990). Substitute fuels for road transport. A technology assessment. OECD/IEA, Paris.

KEMPE'S (1992). Kempe's Engineers Yearbook 1992. Morgan–Grampian, London.

LANGLEY K F (1987). A ranking of synthetic fuel options for road transport applications in the United Kingdom. Department of Energy Paper R–33, Energy Technology Support Unit, Harwell, Oxfordshire.

MARKUS H (1988). "Projektleitung Biologie Okologie und Energie". Proceedings of the "Ecofuel" workshop, Milan, Italy, 8 March 1988.

MARROW J E et al (1987). An assessment of bio–ethanol as a transport fuel in the United Kingdom. Department of Energy Paper R–44, Energy Technology Support Unit, Harwell, Oxfordshire.

MARTIN D J and R A W SHOCK (1989). Energy use and energy efficiency in UK transport up to the year 2010. Department of Energy, Energy Technology Support Unit. London: HMSO.

MICHAELIS L (1991). Global warming impacts of transport. Proceedings of the International Symposium "Transport and Air Pollution", 10–13 September 1991, Avignon, France.

NEW ENERGY DEVELOPMENT ORGANISATION (1987). Technology assessment of various coal–fuel options. NEDO Nuclear Research Centre, Karlsruhe, Germany.

SINGH M K, C L SARICKS, S J LABELLE and D O MOSES (1987). Emerging environmental constraints on the use of gasoline and diesel fuel and tradeoffs associated with the use of alternative fuels. Center for Transportation Research, Argonne National Laboratory, United States of America.

STONE AND WEBSTER ENGINEERING CORPORATION (1985). Economic feasibility study of a wood gasification based methanol plant. Stone and Webster Engineering Corporation, Boston, Massachusetts, United States of America.

TRANSPORT AND ROAD RESEARCH LABORATORY (1992). Energy consumption and air pollution in the road transportation sector. Department of Transport, Vehicles and Environment Division, Vehicles Group, TRRL, Crowthorne.

US DEPARTMENT OF ENERGY (1988). Assessment of costs and benefits of flexible and alternative fuel use in the United States transportation sector. Progress Report 1: Context and analytical framework. DOE.

VOLKSWAGEN AG (Undated). VW Documentation: Research for the future. Alternative fuels. VW Research and Development and Public Relations, Wolfsburg, Germany.

WAGNER T O et al (1987). Comparative economics of methanol and gasoline. SAE Technical Paper Series No 872061, Society of Automotive Engineers, Inc., Warrendale, Pennsylvania, United States of America.

WATERS M H L (1992). UK road transport's contribution to greenhouse gases: A review of TRRL and other research. Transport and Road Research Laboratory Contractor Report 223. Department of Transport, TRRL, Crowthorne.

WORLD METHANOL CONFERENCE (1987). Proceedings. December 1987.

11. HEALTH AND ENVIRONMENTAL EFFECTS OF EXHAUST EMISSIONS

11.1 Introduction

The complete combustion of a carbon–based fuel in oxygen yields carbon dioxide (CO_2) and water. Combustion in the internal combustion engine (ICE) is rarely complete due to the very short time periods available, and it is done in air rather than pure oxygen. The complex exhaust emissions that are formed contribute both directly and indirectly (via subsequent atmospheric chemical reactions) to air pollution, both locally and globally.

Changes in the composition of the atmosphere on a local level can lead to certain health problems, such as caused by the emissions of carbon monoxide (CO), particulates and the generation of ozone. On a global level, the increase of concentration of greenhouse gases is thought likely to create an enhanced global warming with yet unpredictable consequences for the environment and human life. Several of the exhaust emission compound classes are discussed in the following sections.

11.2 Hydrocarbons (HC)

Into this category fall the gaseous products of incomplete combustion plus all the components of the fuel itself that can vaporise. Almost 400 individual organic compounds have been identified in vehicle exhaust (Ball *et al*, 1991, quoting Graedel *et al*, 1986). They represent most of the major classes of organic compounds.

The role of HCs in atmospheric chemistry is very important. The rate of formation of photochemical oxidants is closely related to the rate at which hydrocarbons are scavenged by hydroxyl (OH) radicals, since it is this scavenging that produces organic peroxy radicals which subsequently produce ozone and other oxidants through the oxidation of NO to NO_2. The US has shown that as the science of ozone–forming potential of exhaust pollutants has advanced, the classification of HC has become less useful as an indicator of the emissions' ozone–forming tendency. More descriptive classifications follow.

11.3 Non–methane hydrocarbons (NMHC)

As recognition grew that methane, one of the components of HC, contributes minimally to short–term ozone formation, US regulators created this category with the methane content removed from the organic gases, so that the amount remaining would be more directly relevant to the short–term ozone–forming tendency of the mixture.

11.4 Non–methane organic gases (NMOG)

The NMHC category, taken literally, does not include oxygenated species such as aldehydes, alcohols, ethers and ketones. The NMOG category includes them. Such species are present in both fuels and exhaust emissions. The term "gas" is used to indicate that the component must be in the air to be of significance as a contributor to ozone formation. NMOG is a more precise definition of the carbon–containing emission products of significance to ozone formation than NMHC.

11.5   Volatile organic compounds (VOC)

Table 28 shows estimates of the absolute (k tonnes) and relative contribution of road transport sources to VOC emissions in the UK in 1983 (from Ball *et al*, 1991).

Pollutant	Petrol exhaust	Diesel exhaust	Evaporation of petrol	% from transport
methane	33.0	0.8	0.0	9.2
ethane	3.14*	0.0	0.0	14.5
propane	0.68*	0.0	0.0	12.6
n–butane	33.1	0.0	18.0	64.9
i–butane	14.5	0.0	3.6	55.2
n–pentane	25.2	0.0	6.7	24.2
i–pentane	44.7	0.0	58.1	46.9
ethylene	33.4	5.2	0.0	63.3
propylene	12.0	3.4	0.0	46.2
acetylene	31.0	4.4	0.0	79.9
toluene	61.8	0.0	2.0	42.8
o–xylene	18.0	0.0	0.0	18.1
m–xylene	21.2	0.0	0.0	20.6
p–xylene	15.9	0.0	0.0	16.3
ethylbenzene	18.0	0.0	0.0	18.1
formaldehyde	2.1	15.4	0.0	36.1
acetaldehyde	0.5	3.7	0.0	35.6
propionaldehyde	0.8	0.9	0.0	44.7
butyraldehyde	0.5	0.0	0.0	33.3
i–butyraldehyde	0.0	1.2	0.0	32.4
valeraldehyde	0.1	1.0	0.0	35.5
acetone	1.1	0.9	0.0	6.2
1–butene	0.5	1.9	2.7	24.8
2–butene	0.3	0.0	5.2	33.1
2–pentene	0.7	0.0	2.3	51.7
1–pentene	0.7	1.2	4.3	32.1
2–methyl–1–butene	0.7	0.0	4.3	42.0
3–methyl–1–butene	0.7	0.0	0.8	53.6
2–methyl–2–butene	1.4	0.0	6.5	41.8
butylene	0.5	0.0	0.0	100
benzaldehyde	1.6	0.4	0.0	95.2

* from Bailey *et al* (1990). Except where noted the source was PORG (1987)

Table 28.   Contribution of road transport sources to VOC emissions in the UK

VOCs comprise a very wide range of individual substances, including hydrocarbons (alkenes, alkanes and aromatics), halocarbons (such as trichloroethylene) and oxygenates (alcohols, aldehydes and ketones). All are organic compounds of carbon and are of sufficient volatility to exist as vapour in the atmosphere.

Most measurements of total VOCs are in terms of their carbon content, without analysis as individual compounds. The major contributor to VOCs is normally methane which has a global background concentration of 1.6 ppm (approximately 1,100 $\mu g/m^3$). Whilst most other individual compounds (for example, benzene) are present in urban air at concentrations of a few $\mu g/m^3$, or less, *total* VOCs amount to several hundred $\mu g/m^3$ in concentration in excess of the methane level.

Immense variations occur in the atmospheric behaviour between compounds, with some being highly reactive and of short atmospheric lifetime, and others very long–lived. The short–lived compounds, especially the chemically–reactive HCs, contribute substantially to atmospheric photochemical reactions and thus to formation of ozone, peroxyacetyl nitrate (PAN) and other oxidants. Less reactive substances are dispersed away from urban areas and decompose slowly.

No generalisation can be made with reference to health effects since they are compound–specific. Some VOCs are of significant toxicity and are given detailed consideration in the following text. A number of VOCs are proved or suspected carcinogens.

Many VOCs are of significance in relation to their environmental effects, particularly their contribution to secondary pollution formation and to stratospheric ozone depletion. VOCs also contribute indirectly to formation of atmospheric acidity. Ethylene, a major VOC component, is a plant hormone and can seriously inhibit the growth of plants.

11.6 Methane, ethane and propane

Small quantities of methane and propane are products of incomplete combustion of petrol. When methane, ethane or propane is the fuel, a portion flows through unburnt into the exhaust and then becomes the major component of the emission products.

In the atmosphere methane and ethane are slowly attacked by the hydroxyl (OH) radical. Ethane is scavenged at a rate some 30 times faster than the attack on methane. Formaldehyde or carbon dioxide are likely final oxidation products of methane and acetaldehyde the likely product of ethane (Ball *et al*, 1991). Currently methane is viewed as being photochemically non–reactive over the short term and thus not contributing to ozone formation. However, over a longer term (more than one day) methane does begin to take part in ozone formation, so this view is being reconsidered. Ethane and propane are slightly reactive.

Methane is a greenhouse gas. Because of its low reactivity, it has a long residence time in the atmosphere (3.6 years) compared with most ozone–forming gases but a relatively short residence time compared with other greenhouse gases such as CO_2. On a mass basis, it has a greater radiative forcing effect than CO_2 with expert opinion varying on its contribution. Relative to CO_2, the current estimates of the ratio are somewhere between 10 and 30 to one – the IPCC (1992) estimate of the global warming potential (GWP) of methane is presently 11.

Methane concentrations have shown a gradual rise with time and are now approximately 1.7 ppm with urban concentrations between 1.7 and 4 ppm. As far as toxic effects of methane, ethane and propane are concerned, they are simple asphyxiants.

11.7   Methanol and ethanol

As with natural gas and propane, a portion of the methanol or ethanol entering an engine as fuel is exhausted unburnt and becomes the major component of the exhaust gases. Methanol and ethanol are also found in the exhaust gases from petrol–engined vehicles.

Methanol is mildly toxic by inhalation. When both inhaled and ingested it can cause human systemic effects including headache, vision deterioration, cough, respiratory effects, nausea and vomiting. Its main toxic effect is exerted on the nervous system, particularly the optic nerve and possibly the retina which can progress to permanent blindness.

Compared with methanol, the toxicological effects of ethanol are negligible, although excessive consumption may cause sclerosis of the liver.

11.8   Aldehydes

Aldehydes are formed as products of incomplete combustion in engines using petrol, diesel, methanol, ethanol, propane or methane as fuels. Generally, the presence of methanol or methyl ethers in the fuel will lead to formaldehyde as the primary aldehyde in the exhaust, while ethanol or ethyl ethers will lead to acetaldehyde as the primary aldehyde in the exhaust gases. In both cases, other aldehydes are present, but in much smaller quantities.

In the atmosphere, aldehydes are among the more reactive contributors to ozone formation. Under photochemical conditions formaldehyde may represent the major source of hydrogen atoms that then become available for subsequent involvement in atmospheric photochemistry (Ball *et al*, 1991, quoting Finlayson–Pitts and Pitts, 1986). Formaldehyde is also formed from other reactive organic compounds in the atmosphere due to photochemical reactions. Current evidence in the US indicates that increased formaldehyde exhaust emissions do not lead to increased atmospheric formaldehyde levels, except perhaps in enclosed spaces such as garages and tunnels.

Formaldehyde is a natural constituent of the air present in variable concentrations that are typically a few micrograms per cubic metre. Levels of 10–15 ppb or 0.01–0.015 ppm, were measured in air near to roads in Switzerland and 0.1–2.1 ppb at Harwell, Oxfordshire, in the summer of 1974. However, in urban air much higher concentrations are found and 10 ppb (12.3 μg/m^3) is typical with almost 100 ppb expected in bad photochemical smogs.

There are a number of international recommendations for maximum exposure levels, mainly relating to occupational exposure. A recommendation of the World Health Organisation (1987), that the concentration of formaldehyde in non–industrial buildings should not exceed 0.1 mg/m^3 (about 0.1 ppm) is almost an order of magnitude higher than some reported ambient concentrations.

While automotive emissions appear to make a substantial contribution to the concentration of formaldehyde in urban air, it is important to consider other major sources. The highest concentrations most people encounter are indoors rather than outdoors – in buildings where chipboard has been used as a construction material, values can exceed 1 mg/m^3. Although better building material product control has meant that such high values are now less common, formaldehyde concentrations in conventional buildings can still range between 50 and 100 μg/m^3. Smoking also represents a source of formaldehyde and this would amount to about a milligram for a person smoking 20 cigarettes.

Formaldehyde and acetaldehyde are toxic and possibly carcinogenic. They are specifically named in the 1990 US Clean Air Act Amendments (CAAA) as to be controlled as toxic air pollutants. Formaldehyde is a poison by inhalation, ingestion and skin contact. Acetaldehyde is a poison by ingestion and an irritant by inhalation and skin contact.

Studies of the toxicological effects of inhaled formaldehyde show that acute exposures can kill experimental animals, while somewhat more moderate exposures (12–50 mg/m^3) indicate a range of respiratory changes. At even lower exposures there are marked differences in the irritancy of formaldehyde.

In humans formaldehyde odour can be detected at about 0.1 mg/m^3 with eye and throat irritation at concentrations about a factor of five above this. Concentrations of around 5 mg/m^3 prove tolerable for about thirty minutes. Once the concentrations drop below about 2.4 mg/m^3 changes begin to be difficult to detect and it seems unlikely that chronic obstructive lung disease would occur in people exposed to concentrations below 1.8 mg/m^3 (WHO, 1987).

There has been considerable interest and controversy over the carcinogenic effects of formaldehyde (Graham *et al.*, 1988). It has been shown to be carcinogenic in several in vitro systems. In rats at high doses (18.7 mg/m^3) a high incidence of squamous–cell carcinomas is found, but the dose response relationship is highly non–linear. Human studies provide no conclusive evidence of carcinogenicity, although their statistical reliability is low (WHO, 1987).

The WHO guideline value given above might also be used for estimates of the nuisance arising from formaldehyde in the roadside environment, but it is unlikely that concentrations as high as 0.1 mg/m^3 will be frequent. Given the fact that most of our exposure to formaldehyde is not from outdoor atmospheric sources, even the possibility that it is a carcinogen does not seem to make it particularly relevant in the near field. However, public perception can mean that even possible carcinogens can become a sensitive issue.

11.9   Olefins

Olefins are highly reactive unsaturated compounds. This means that there are carbon atoms in the molecule that are able to accept additional atoms such as hydrogen or chlorine, and it is this attribute which tends to make them highly reactive in ozone-forming reactions and toxic to living systems.

Some are present in petrol as a result of refinery manufacturing processes such as cracking. Some are created in the engine during combustion, although most can be removed by the catalytic converter. They tend to be ozone formers and tend to be toxic. They generally have a moderate energy content and are desirable fuel components.

One specific olefin, 1,3–butadiene, is named in the CAAA to be controlled as a toxic air pollutant and is believed to be the most toxic of those named.

11.10 Aromatics

Aromatics are carbon compounds with the carbon atoms strung together in rings. The basic ring has six carbon atoms and is shaped like a regular hexagon. Some heavier aromatics with two or more hexagonal rings with common sides (polycyclic aromatic hydrocarbons – PAHs) are also present in petrol and some are formed as a result of pyrolytic (thermal degradation) during combustion. Aromatics are more resistant than olefins to removal in the catalytic converter. Aromatics have a high energy content which makes them desirable fuel components.

Some aromatics are ozone–forming and some are toxic. Benzene and PAHs are named as toxics in the CAAA. Xylenes and some of the more complex aromatics are active ozone formers with an atmospheric residence time of about a day.

Benzene is a colourless liquid that is fairly stable chemically, but is highly volatile. Ambient concentrations of benzene are typically between 3 and 160 μg/m^3. In London the annual mean is 10–12 μg/m^3. Levels close to major emission sources (for example, petrol stations) can be as high as several hundred $\mu g/m^3$. About 80 percent of anthropogenic emissions of benzene comes from petrol–fuelled cars. This results from both the benzene content of the petrol and the pyrolysis (thermal breakdown) of petrol. A further 5 percent of benzene emissions comes from the handling, distribution and storage of petrol, with about one percent coming from oil refining.

The literature on the health implications of benzene is vast, as are the controversies surrounding its toxicology (Graham *et al*, 1988). At levels of occupational exposure (several hundred mg/m^3) there is a clear excess incidence of leukaemia. However, no case of leukaemia has been confirmed following regular and repeated occupational exposure to benzene in air at concentrations below 320 mg/m^3. Early manifestations of toxicity include anaemia, and persistent exposure to toxic levels of benzene may cause injury to the bone marrow. No adverse effect on blood formation in humans has been confirmed following regular and repeated occupational exposure to benzene in air concentrations below 80–96 mg/m^3.

As a known human carcinogen, benzene is placed in the IARC (International Agency for Research on Cancer) Group 1. Much of the data for the carcinogenicity of benzene comes from exposed populations of workers. The IARC carcinogenicity categories are shown in Table 29 in section **11.17.**

A WHO (1987) report gives a figure of $4x10^{-6}$ lifetime risk of leukaemia from a continual exposure to 1 $\mu g/m^3$. There are particular problems in setting limits for a carcinogen when it can be argued that there is no threshold below which harm no longer occurs and they consider that no safe level can be recommended as benzene is a carcinogen.

It is not easily possible to develop quantitative risk assessment methods for considering roadside carcinogens but it is worth noting the kind of risk that might be imposed on an individual at the edge by a busy road. Bailey *et al* (1990) suggest annual mean concentrations for benzene at the kerbside in London to be 28–31 $\mu g/m^3$. Hourly concentrations could be much higher. It would be possible for a person walking along a busy road during the rush hour to be exposed perhaps to 60 $\mu g/m^3$ for an hour each day. This would present a lifetime risk of $1x10^{-5}$. This might be compared with continual exposure to 10 $\mu g/m^3$ within a house which would yield a lifetime risk of $4x10^{-5}$.

The best studied and most–measured PAH is benzo[a]pyrene (BaP). BaP concentrations in most cities are in the range 1–5 ng/m^3, rising to perhaps 10 ng/m^3 in major urban areas, although values of the order of 100 ng/m^3 were recorded in the 1960s when the use of open coal fires for domestic heating was more common.

Some PAHs are useful indicators of sources. Coronene, for example, is particularly prevalent in motor vehicle exhausts, and present in only small quantities in coal smoke.

It is generally argued, on the basis of skin painting experiments, that benzo[a]pyrene and dibenz[a,h]anthracene are the most carcinogenic PAH compounds frequently found in ambient air. The benzo[a]fluoranthenes are regarded as moderately carcinogenic with benzo[a]anthracene and chrysene rather weaker. High levels of lung cancer have been observed in coke oven workers exposed to a BaP concentration of about 30 ng/m^3. There are no known toxic effects other than carcinogenicity.

11.11 Nitrogen oxides (NO$_X$)

Oxides of nitrogen are formed by reaction of atmospheric oxygen and nitrogen (and any nitrogen in the fuel) at the high temperatures and pressures found within the engine's combustion chamber. The higher the temperature, the greater the amount of NO$_X$ in the exhaust gases. There are several different oxides, with nitric oxide (NO) the major constituent. Some Nitrogen dioxide (NO$_2$) is also present in the exhaust and rapidly forms in the atmosphere as the exhaust NO is oxidised. Nitrous oxide (N$_2$O), a greenhouse gas, is also present at low levels, although the effect of three–way catalytic converters can increase its concentration in the atmosphere due to the catalytic reduction of the higher oxides. Globally, quantities of nitrogen oxides produced naturally (by bacterial and volcanic action and by lightning) far outweigh anthropogenic emissions.

NO$_2$ is generally regarded as being most important from the point of view of human health. Consequently, data on health risks, ambient concentrations and standards and guidelines are generally expressed in terms of NO$_2$ rather than NO$_X$. NO$_2$ is a reddish-brown gas that is a strong oxidant and soluble in water. Annual mean concentrations in urban areas are generally in the range 20–90 µg/m^3. Levels vary significantly throughout the day, with peaks generally occurring twice daily as a consequence of "rush–hour" traffic. Maximum daily and half–hour means can reach 400 µg/m^3 and 850 µg/m^3 respectively.

NO$_2$ is an experimental poison and moderately toxic to humans by inhalation. A variety of respiratory system effects have been reported to be associated with exposure to short and long–term NO$_2$ concentrations less than 3.8 mg/m^3 in humans, including altered lung function and increased prevalence of acute respiratory illness. Animal toxicology studies have shown lung tissue damage, development of lesions in the lung, and increased susceptibility to infection. Certain human health effects may occur as a result of exposures to NO$_2$ concentrations at or approaching ambient levels.

Human pulmonary function effects resulting from single, short–term exposures of less than three hours duration have been unambiguously demonstrated only at concentrations (greater than 1.9 mg/m^3) well in excess of ambient exposure levels typically encountered by the public. More subtle health effects that were of uncertain health significance, such as mild symptomatic effects, had been reported for some asthmatics after a single two-hour exposure to about 1 mg/m^3. Young children are also considered at greater risk from ambient NO_2 exposures.

The European Community has set a limit value of 200 µg/m^3 for the 98th percentile of one hour average concentrations during a calendar year (EC Directive 85/203/EEC, Official Journal of the European Communities, 1985). Other standards include the US National Ambient Air Quality Standard (NAAQS) that the annual mean concentration should not exceed 100 µg/m^3 (52 ppb) and the recommendations of the World Health Organisation for maximum one–hour and 24–hour average concentrations of 400 µg/m^3 (210 ppb) and 150 µg/m^3 (80 ppb) respectively. Though these numbers and the exposure periods to which they refer vary considerably, Ball *et al* (1991) demonstrated that their severity was more or less equivalent.

Environmental effects of NO_2 and NO_X compounds include increased acidic deposition and vegetation effects. Visible injury to vegetation due to NO_2 alone occurs at levels which are above ambient concentrations, except around a few point sources. Nitrogen oxides are critical to ozone formation. When NO_2 is subjected to ultraviolet radiation from sunlight, an oxygen atom may be removed from the molecule. If this oxygen atom unites with a molecule of oxygen (O_2), ozone (O_3) is formed.

11.12 Carbon monoxide (CO)

Carbon monoxide is a colourless, odourless, tasteless gas that is slightly less dense than air. It is a relatively stable compound and thus takes part only slowly in atmospheric chemical reactions. Natural background levels of CO fall in the range 0.01–0.23 mg/m^3. Levels in urban areas are highly variable, depending upon weather conditions and traffic density. Eight–hour mean values are generally less than 20 mg/m^3 but can be as high as 60 mg/m^3 (CEC, 1991).

Carbon monoxide is an intermediate product through which all carbon species must pass when oxidised. In the presence of adequate O_2, most CO produced during combustion is immediately oxidised to CO_2, but this is not the case during combustion in the ICE, especially under idling and deceleration (overrun) conditions. Carbon monoxide is a pollutant in its own right because of its direct health effects.

When CO is inhaled it can enter the blood stream and disrupt the supply of O_2 to body tissues. The health effects of CO result principally from its ability to displace O_2 on haemoglobin, forming carboxyhaemoglobin (COHb). The consequent reduced oxygen availability (hypoxia) can give rise to a wide range of health effects depending on how much the flow of O_2 to the body is impeded.

Health effects are usually related to blood levels of COHb (expressed as a percentage), which can in turn be related to exposure (as a function of exposure time as well as concentration). The "no-observed-effects" level is about 2 percent COHb which can be related to an eight-hour exposure (of moderate human activity) to 15-20 mg/m^3. Certain neurobehavioural effects can be expected at 5 percent COHb (moderate activity for eight hours in 40 mg/m^3) that can include impaired learning ability, reduced vigilance, decreased manual dexterity, impaired performance of complex tasks and disturbed sleep.

The World Health Organisation (1987) recommended a COHb level of 2.5-3 percent for the protection of the general public and suggested the following guidelines for limiting exposure to CO in order to maintain COHb below that level: a maximum exposure of 100 mg/m^3 (85 ppm) for periods not exceeding 15 minutes and time weighted average exposures at the following levels for the exposure periods indicated: 60 mg/m^3 (50 ppm) for 30 minutes, 30 mg/m^3 (25 ppm) for one hour, 10 mg/m^3 (10 ppm) for eight hours.

In addition, increased risk of certain effects on the cardiovascular system can be expected to begin at levels close to peak ambient conditions. These effects include reduced exercise and physical work capacity, enhanced development of coronary heart disease and even contribute to heart attack. Elevated COHb levels can also reduce the availability of O_2 to the central nervous system (CNS) including the brain. High levels may cause strokes and repeated episodes of impaired O_2 supply would be expected to damage the ability of the CNS to transmit information. Individuals most at risk to the effects of CO include those with existing cardiovascular or chronic respiratory problems, the elderly, young children and foetuses.

Carbon monoxide is also thought to participate in a number of other atmospheric reactions, including ozone formation, where its effect is of lower direct significance than NMOGs and NO_x. The main atmospheric reaction in which the CO is oxidised is with the hydroxyl (OH) radical. For that reason, CO also acts indirectly as a greenhouse gas by consuming OH radicals that would otherwise promote removal of more powerful greenhouse gases, primarily methane. Plants both produce and metabolize CO and are only harmed by prolonged exposure to very high levels. The lowest levels of which significant effects on vegetation have been reported is 115 mg/m^3 for 3 to 35 days.

11.13 Particulates

Particulate matter is a mixture of organic and inorganic substances, present in the atmosphere in both liquid and solid form. Coarse particles can be regarded as those with a diameter greater than 2.5 μm, and fine particles less than 2.5 μm. A further distinction that can be made is to classify particulates as either "primary" or "secondary", according to their origin. Primary particulate matter from the ICE comprises soot-like particles formed by incomplete combustion of fuel. The particles are primarily carbon, but adsorbed on their surface are other products of incomplete combustion such as polycyclic aromatics. Not only are they physically dirty, but they are suspected as being carcinogenic.

Particulate emissions from the combustion of diesel fuel are much greater than from petrol combustion. Van den Hout and Rijkeboer (1986) reported emission factors of 4 to 7 g/litre of fuel for light duty diesel vehicles and 7 to 14 g/litre for heavy duty diesels, compared with 0.65 g/litre for petrol vehicles (6 to 22 times as much). When expressed in the more common units of grammes per kilometre, the ratio of emissions of particulates from diesel and petrol vehicles might be typically 500:1, primarily because in the UK, most diesel is used by heavy duty vehicles whose rate of fuel consumption is much higher than that of petrol cars.

Reported concentrations vary according to the sampling technique. In urban areas, typical annual mean values are 50–150 $\mu g/m^3$ by the gravimetric method. Peak values are 200–400 $\mu g/m^3$. Apart from the ICE, particulate matter is emitted from a wide range of sources including power stations, domestic coal burning and industrial incinerators. Natural sources are less widespread, such as volcanoes and dust storms.

Short–term health effects of exposure to particulates and black smoke include increased mortality and deficits in pulmonary function. Some of the "lowest–observed effect" levels for short–term exposure to particulate matter are: excess mortality – 500 $\mu g/m^3$ (smoke); increased acute respiratory morbidity (adults) – 250 $\mu g/m^3$ (smoke) and decrements in lung function (children) – 180 $\mu g/m^3$ (total suspended particulates)(CEC, 1991). Smoke levels of up to 1,500 $\mu g/m^3$ occurred in the 1952 London smog but, in addition, exposure to air pollutants and especially particulates may give rise to feelings of discomfort, which may cause annoyance.

There is still considerable uncertainty over the significance of the organic fraction of particulates, and especially on the carcinogenic effects of PAHs, as carcinogenic agents in people exposed to normal levels of vehicle pollution. Waller (1987) reviewed the health aspects of diesel engine emissions and the following comments are taken from his conclusions:

"The particular concern expressed about diesel engines in relation to health is over the smoke emissions and their possible carcinogenic activity. The recent animal inhalation studies lend support to this in principle though they do not have any direct bearing on the practical problems ...... even ignoring the uncertainties of extrapolation from animals to man, all human experience would fall far short of the cumulative doses received by animals even in low dose groups in the experimental inhalation studies, in which there were no positive results."

Other environmental effects include the soiling of exposed surfaces, impairment of visibility, potential modification of climate and contribution to acid deposition. About 70 percent of the soiling caused by air pollutants in the UK is from diesel smoke. The situation in towns is more extreme since a greater proportion of the total smoke, and therefore of soiling, is from traffic. Ball and Caswell (1983) estimated that two–thirds of smoke in London was from road traffic; national statistics suggest that the proportion may now be approximately 85 percent which would cause 95 percent of the soiling. Little is known in detail of the effects of particulates on atmospheric chemistry. They absorb and scatter radiation directly and are important in the formation of clouds, but the net effect is not known. Indeed, it is not even known whether the effect of particles in the atmosphere is to increase or decrease the greenhouse effect.

11.14 Lead

Lead is a bluish or silver–grey soft metal. In the context of air pollution, two of its most important compounds are tetraethyl and tetramethyl lead, which are used extensively as "anti knock" additives in petrol.

Historically, the principal source of atmospheric lead has been the combustion of alkyl lead additives in motor fuels. However, the contribution from this source is decreasing in most countries as a result of controls on the lead content of fuels and the availability of lead free petrol. In rural areas lead levels fall in the range 0.1–0.3 $\mu g/m^3$ and most European urban values are now less than 1 $\mu g/m^3$. This compares with roadside concentrations of lead in London during the 1970s of between 0.5 to 3 $\mu g/m^3$ (Schwar and Ball, 1983).

Human exposure to lead is through inhalation of airborne lead and ingestion of lead in foodstuffs and beverages. Blood lead concentrations are a good indicator of recent exposure to lead from all sources, and adverse health effects tend to increase in severity with increasing blood lead level. US Environmental Protection Agency (EPA) standards are based upon the concept of a relationship between ambient air lead and blood lead concentrations, and presume that a blood lead level of 0.15 $\mu g/ml$ (mean value for children) can be achieved at an ambient air lead level of 1.5 $\mu g/m^3$.

The most sensitive body systems to the effects of lead are the nervous and renal systems. In addition, lead has been shown to affect the normal functions of the reproductive, endocrine, hepatic, cardiovascular, immulogic and gastrointestinal systems. The most sensitive group to lead poisoning is children. Some studies indicate that children with high levels of lead accumulated in their baby teeth experience more behavioral problems, lower IQs and decreased ability to concentrate, but conclusions from some studies were often conflicting.

In 1972 the Secretary of State for the Environment implemented a phased reduction in the maximum lead content of petrol, then set voluntarily at 0.84 g/l. This policy was followed between 1972 and 1978 on the basis that the usage of lead additives was maintained at a constant level while the total usage of petrol increased. In 1981 the maximum lead content was further reduced to 0.40 g/l, in compliance with EC Directive 78/611/EEC (Official Journal of the European Communities, 1978), which required member states to set maximum levels in the range 0.15–0.40 g/l. In 1986, due to recommendations from a working party of the UK Department of Health (Lawther, 1980) that the emission of lead to the air from road traffic and other sources should be progressively reduced, the Government announced that the maximum lead content of petrol would be reduced to 0.15 g/l.

Lead is generally toxic to both plants and animals, and although no serious effects are generally seen at current environmental levels, it is widely considered prudent to limit further dispersal of lead as far as is possible. An air quality standard for lead has been set for the European Community at 2 µg/m^3 as an annual mean concentration not to be exceeded (EC Directive 82/884/EEC, Official Journal of the European Communities, 1982). Other standards and recommendations include those of the USA, that the three month average concentration should not exceed 1.5 µg/m^3 (Federal Register, 1978) and of the World Health Organisation, whose guideline recommendation is that the long–term average concentration should not exceed 0.5 to 1.0 µg/m^3 (WHO, 1987).

11.15 Ozone (O_3) and PAN

Ozone (O_3) is the tri–atomic form of molecular oxygen. It is one of the strongest oxidising agents, which makes it highly reactive.

Peroxyacetyl nitrate (PAN) is an oxidising agent formed by the reaction of organic compounds (for example, aldehydes) with OH radicals, followed by the addition of O_2 and NO_2.

Most of the ozone in the troposphere (lower atmosphere) is formed indirectly by the action of sunlight on nitrogen dioxide. About 10–15 percent of tropospheric ozone is transported from the stratosphere where it is formed by the action of UV radiation on oxygen. In addition to ozone, photochemical reactions produce a number of oxidants including PAN, nitric acid and hydrogen peroxide, as well as secondary aldehydes, formic acid, fine particulates and an array of short–lived radicals. As a result of the various reactions that take place in the atmosphere, ozone tends to build up downwind of urban centres (where most of the NO_x is emitted from vehicles).

Background levels of ozone in Europe are usually less than $30\mu g/m^3$ but can be as high as $120 \ \mu g/m^3$. Maximum hourly values may exceed $300 \ \mu g/m^3$ in rural areas and $350 \ \mu g/m^3$ in urban areas. PAN concentrations seem to have a diurnal pattern similar to that of ozone, with the peak value of PAN occurring several hours before that of ozone. Daily mean values for PAN tend to be some 2–20 percent of values for ozone. Maximum reported values for PAN approach $100 \ \mu g/m^3$.

Ozone and other oxidants can cause a range of acute effects including eye, nose and throat irritation, chest discomfort, cough and headache. These have been associated with hourly oxidant levels of about $200 \ \mu g/m^3$. Pulmonary function decrements in children and young adults have been reported at hourly average concentrations in the range $160–300 \ \mu g/m^3$. Increased incidence of asthmatic attacks and respiratory symptoms have been observed in asthmatics exposed to similar levels of ozone. The non–ozone components of the photochemical mixture cause eye irritation at ozone levels of about $200 \ \mu g/m^3$.

Other environmental effects include damage to materials (including as a result of prolonged exposure to low concentrations) and vegetation effects (some claims have been put forward that high ozone concentrations cause a reduction in plant growth).

11.16 Carbon dioxide (CO_2)

CO_2 (along with water vapour) is the major product of carbon–fuel combustion. It plays no significant role in ozone formation and is not toxic. It is a greenhouse gas since it absorbs heat energy radiating from the earth's surface. It is agreed that of the various greenhouse gases, CO_2 contributes to around 50 percent of the warming effect. It is used as the basis for radiative forcing comparison with an assumed factor of one.

11.17 IARC Categories

Reference has been made at times in this chapter to the likelihood of various compounds being carcinogenic, and the IARC (International Agency for Research on Cancer) category quoted. Ball *et al* (1991) report the IARC carcinogenicity categories and these are repeated in Table 29 overleaf.

Group	Category description
1	Proven human carcinogen. This category includes chemicals or groups of chemicals for which there is sufficient evidence from epidemiological studies to support a casual association between exposure and cancer.
2	Probable human carcinogen. This category includes chemicals and groups of chemicals for which, at one extreme, the evidence of human carcinogenicity is almost sufficient, and those for which, at the other extreme, it is inadequate. To reflect this range, the category is divided into two subgroups according to higher (Group 2A) and lower (Group 2B) degrees of evidence.
2A	This group is usually for chemicals for which there is at least limited evidence of carcinogenicity in humans and sufficient evidence for carcinogenicity in animals.
2B	This group is usually used for chemicals for which there is inadequate evidence of carcinogenicity in humans and sufficient evidence of carcinogenicity in animals. In some cases the known chemical properties of a compound and the results of short–term tests have allowed its transfer from Group 3 to Group 2B, or from Group 2B to Group 2A.
3	Unclassified chemicals. This group includes chemicals or groups of chemicals which cannot be classified as to their carcinogenicity in humans.

Table 29.   IARC carcinogenicity categories

REFERENCES FOR CHAPTER 11

BAILEY J C, B SCHMIDL and M L WILLIAMS (1990). Speciated hydrocarbon emissions from vehicles operated over the normal speed range on the road. *Atmos. Environ.*, **24A**, 43–52.

BALL D J and R CASWELL (1983). Smoke from diesel–engined road vehicles: an investigation into the basis of British and European emission standards. *Atmos. Environ.* **17**, 169–181.

BALL D J, P BRIMBLECOMBE and F M NICHOLAS (1991). Review of air quality criteria for the assessment of near–field impacts of road transport. TRRL Contractor Report 240. Transport and Road Research Laboratory, Crowthorne.

COMMISSION OF THE EUROPEAN COMMUNITIES (1991). Research and Technology Strategy to Help Overcome Environmental Problems in Relation to Transport (SAST Study No 3). DG XII: Science, Research and Development: Study 1: Draft Final Report, October 1991.

FEDERAL REGISTER (1978). National primary and secondary ambient air quality standard for lead. US EPA Federal Register 43, FR46246.

FINLAYSON–PITTS B J and J N PITTS (1986). Atmospheric chemistry. John Wiley, New York.

GRAEDEL T E, D T HAWKINS and L D CLAXON (1986). Atmospheric chemical compounds. Academic Press, Orlando, Florida.

GRAHAM J D, L C GREEN and M J ROBERTS (1988). In Search of Safety. Harvard University Press.

GUSHEE D E (1992). Alternative fuels for automobiles: Are they cleaner than gasoline? Congressional Research Service Report for Congress. Report 92–235 S, 27 February 1992, The Library of Congress, Washington, DC, United States of America.

INTERGOVERNMENTAL PANEL ON CLIMATE CHANGE (1992). 1992 IPCC Supplement. World Meteorological Organisation/United Nations Environment Programme, Cambridge University Press, Cambridge.

LAWTHER (1980). Lead and Health. The report of a DHSS working party on lead in the environment. HMSO.

OJEC (1978). Council Directive of 22 July 1978 for the lead content of petrol, 78/611/EEC. Official Journal of the European Communities, 19–21, L197.

OJEC (1982). Council Directive of 3 December 1982 on a limit and guide value for lead in air, 82,884/EEC. Official Journal of the European Communities, L378.

OJEC (1985). Council directive on an air quality limit and guide value for nitrogen dioxide, 85/203/EEC. Official Journal of the European Communities.

PORG (1987). Ozone in the United Kingdom. The first report of the United Kingdom photochemical oxidants review group. Department of the Environment.

SCHWAR M J R and D J BALL (1983). Thirty years on: A review of air pollution in London. Greater London Council.

VAN DEN HOUT K D and R C RIJKEBOER (1986). Diesel exhaust and air pollution. TNO, publication number R/86/038.

WALLER R E (1987). Diesel engine emissions – health aspects. Evidence taken before the European Communities Committee (Sub–committee F). House of Lords, session 1987–88, 7th report. HL paper 32.

WHO (1987). Air Quality Guidelines for Europe. World Health Organisation, HMSO.

12. SUMMARIES AND CONCLUSIONS

12.1 Introduction

This book presents a discussion of the benefits, disadvantages and prospects for alternative fuels and an appraisal of the future for conventional diesel and petroleum fuels for road transport applications. Implications for local or regional air pollution and global concerns such as an enhanced greenhouse effect, together with indications of the cost of, and infrastructure and technology requirements for, the use of alternative fuels, have been considered and reported. Examples of demonstration alternatively–fuelled vehicle projects have been given and note has been made of the global automotive industry's efforts at developing engines and vehicle systems to accommodate new fuels, especially with regard to complying with increasingly stringent exhaust emission legislation.

The following sections present a brief summary of each alternative fuel, followed at the end by conclusions as to which display greatest promise and are likely to find widescale vehicular application in future.

12.2 Liquid hydrocarbons

Reformulated fuels are appealing since they require no vehicle adjustments (though these may be desirable under certain circumstances to maximise performance) or new infrastructure, apart from modifications to existing refineries. Of particular value is the potential to use reformulated petrol, for example, to reduce emissions from existing vehicles, rather than waiting for fleet turnover or the use of other alternative fuels. This is important since certain vehicle technology classes are more likely to benefit from reformulated fuels than others – uncontrolled petrol engines (which will form the majority of the EC passenger car fleet until about 1996/97 and will still represent 25% in 2002) are likely to demonstrate higher absolute emissions reduction than newer cars with three–way catalysts.

It is probable that Europe will adopt certain fuel reformulations sometime in the future, although not necessarily in line with the present US proposals. Certain changes in fuel specification have already been mandated, such as diesel fuel sulphur content. Due to potential health concerns, the limitation of aromatic content (for petrol and diesel) and the reduction of benzene in petrol is likely, and octane is likely to be maintained by the addition of oxygenates, with a consequent beneficial reduction in exhaust CO and in the quantity and reactivity of HC emissions. Two of the key oxygenates that are promoted for use in petrol are MTBE and ETBE since they have sufficiently low vapour pressures, high octane and a reasonable oxygen content. Ethanol, the main oxygenate competitor to MTBE, is distilled from grain (and can be produced from other biomass) and has advantages as a "renewable" additive, helping to extend conventional fuel supplies.

It is difficult to predict how much reformulated petrol the refineries would be capable of producing. It offers fewer benefits in energy security or greenhouse gas emissions than other alternative fuels because it is mostly oil–based and will increase refinery energy usage. A life cycle analysis should be undertaken in determining how fuels are changed. For example, changing a fuel specification to achieve lower vehicle exhaust emissions may be offset by an increase in processing energy (and emissions produced somewhere else) – such as with diesel fuel desulphurisation.

In addition to lowering the sulphur content of diesel fuel, increasing the cetane rating has been demonstrated as an effective way of reducing exhaust emissions. Other, less well proven means of reducing diesel emissions include reducing the fuel density and distillation range, but given the oil industry's concerns about the implications for reduced supply, this route seems less likely.

In two or three decades crude oil prices are likely to have risen significantly in real terms, especially if global petrochemical demand continues to increase and developing countries aspire to the standard of living and level of personal mobility enjoyed in the West. Other forms of liquid hydrocarbon fuels, derived from unconventional sources, are then more likely to become economical to produce. The most viable form of synthetic fuel able to be introduced is petrol from fuel oil or very heavy crude and, if crude oil prices rise substantially, petrol from coal and kerosine/gasoil from coal.

12.3   Methanol

Methanol's major advantages in vehicular use are that it is a convenient, familiar liquid fuel that can readily be produced from natural gas and other feedstocks using well–proven technology. As a blend of 85 percent methanol/15 percent petrol (M85), it is a fuel for which vehicle manufacturers can, with relative ease, design either a dedicated or flexible fuel vehicle that will outperform an equivalent petrol–fuelled vehicle and obtain an advantage in some combination of emissions reduction and efficiency improvement. Major disadvantages of methanol are the likelihood that it will cost more than petrol, especially during early years of its possible introduction; a loss of up to half the normal vehicle operating range is a consequence of its use, unless a larger fuel tank is installed; and the need for a separate fuel distribution infrastructure. Methanol is more toxic than petrol although its lower flammability may lead to reductions in vehicle fires.

The use of methanol made from natural gas is unlikely to provide any greenhouse benefit and by using coal as the feedstock would significantly increase greenhouse gas emissions. Increased energy security from the widescale use of methanol manufactured from natural gas may only last a few decades unless large new reserves of natural gas are discovered outside of the Middle East OPEC countries.

Proposals in the United States for introducing the vehicular use of methanol in ozone non-attainment areas have been very controversial due to widely varying and competing claims of its expected air quality benefits. While most commentators conclude that methanol has the potential to reduce ozone formation by virtue of emissions of less reactive hydrocarbons, the actual effects are poorly quantified and highly variable, due to location, meteorological conditions and the essentially prototype nature of many available methanol vehicles.

Claims for the expected costs of methanol have ranged from "competitive with and possibly below petrol costs" to "much higher than petrol." Much of the range can be accounted for by differing assumptions as to the scale of the fuel's introduction, likely gas feedstock sources and others. It is believed that prospects for methanol's market success would benefit from improvements in natural gas conversion processes for fuel production, the development of world trade in methanol produced from remote gas sources, firmer evidence of major air quality benefits and the development of more practical cold-starting methods for M100 and improved formaldehyde emission controls.

12.4 Ethanol

Ethanol, like methanol, is a familiar liquid fuel that can be quite easily used, with few problems, in vehicles competitive in performance with petrol-fuelled vehicles. Important advantages are its ease of use as a petrol fuel additive suitable for existing vehicles, and its attractiveness as a stimulus to the agricultural economy.

Greenhouse gas emission benefits are claimed for fuels produced from biomass, but the wide range of estimates of impacts, from virtually no greenhouse impact to one greater than the emissions from using fossil fuels, implies that no firm conclusions can yet be drawn. Ethanol made from food crops (such as corn) appears to be the most expensive and energy-intensive of the major alternative fuels. An important consideration of the use of biofuels is the avoidance of byproduct market saturation which may reduce the environmental benefit claimed for the fuel.

Ethanol's likely contribution to improved air quality has been another area of some contention. A ten percent ethanol/petrol blend (E10 or "gasohol") has demonstrated reduced carbon monoxide emissions, even in newer vehicles. Also, although addition of ethanol to petrol increases its vapour pressure and thus its evaporative emissions, this effect is claimed to be compensated by the emissions' lower reactivity combined with lower CO emissions. The introduction of ethanol as a vehicle fuel would benefit from confirmation of its emissions performance as a neat fuel (E100) in catalyst-equipped vehicles, the development of low-cost production systems using woody biomass as a feedstock, the improvement in distillation technology and the development of an international market in the fermentation byproducts from ethanol production.

12.5   Biodiesel and vegetable oils

Vegetable oils have been demonstrated as technically viable motor fuels in diesel engines, causing few problems if the oil is modified with alcohol to form a methyl ester. Rapeseed methyl ester ("bio–diesel") is used extensively as a fuel in central European farming communities, being an ideal use for surplus rape production. Other niche markets are likely to become established, for example, in developing countries where oil from locally–grown crops can be substituted for scarce or expensive conventional diesel fuel.

Vegetable oil fuels are a renewable energy source, and while experts may argue as to the precise greenhouse impact of their use as fuels, they enable conventional fossil fuel supplies to be extended. Emissions benefits from the use of vegetable oils is still uncertain, with considerable variation depending on vehicle application and its operating cycle. Further development of fuel standards and the formulation of additives are required to ensure adequate performance, engine and fuel component protection and established and well–known emissions characteristics.

12.6   Liquefied petroleum gas

Liquefied petroleum gas is a well established fuel, both in stationary and motor vehicle applications. Around four million vehicles worldwide are fuelled by LPG and expectations are that the global market will grow substantially. One of the biggest growth in LPG usage (and also CNG) will occur in the former Soviet Union and Eastern Europe generally, since the CIS intends using more of its indigenous gas reserves, from which LPG is a byproduct. In the UK LPG has never enjoyed a large usage as a motor fuel, partly due to the expensive infrastructure necessary, vehicle conversion cost and deteriorating cost advantage with respect to petrol and diesel fuels.

Some niche market applications, for example in centrally–fuelled fleets, may appeal to certain operators. It is not thought likely that LPG will extend its vehicle fuel market share in this country – it is more probable that its supply will remain a static proportion of refinery throughput (a byproduct) and that measures to expand production (that are more costly and would upset the mix of other petroleum products) would not be put in place without a major form of incentive for its increased usage.

12.7   Natural gas

Natural gas may be cheaper as a fuel than petrol although the net cost to the consumer depends on the distribution and refuelling system, and the total cost to the motorist rises if vehicle costs are added. Natural gas can fuel a dedicated vehicle of equal performance to petrol–fuelled vehicles, with potentially lower emissions and equal or higher efficiency.

Of particular concern to the United States is the ability of natural gas to demonstrate large ozone benefits much more clearly than is the case with M85. Other important advantages include the availability of an extensive distribution network and experience of gas handling. The use of natural gas may confer a moderate greenhouse benefit, because of methane's low carbon/hydrogen ratio, but the effect is highly sensitive to several system variables such as estimating gas leakage from the distribution system and continuing uncertainties as to the warming impact from methane emissions.

The use of natural gas could enhance energy security especially as gas reserves are better globally dispersed than crude oil. Natural gas in the form currently most used in vehicles – as compressed natural gas (CNG), has some significant drawbacks as a vehicle fuel. These include the requirement for bulky gas cylinders, the low volumetric energy density of the fuel and higher vehicle cost.

In establishing a natural gas vehicle fuel market, transitional arrangements must be made – such as providing dual–fuel vehicles which would have high first costs and may suffer from performance reduction when using gas. Some of the disadvantages, particularly the range limitations, may be somewhat ameliorated by storing gas in its denser form as liquefied natural gas (LNG).

12.8   Electricity

Electricity as a vehicle "fuel" has the important advantage of having an available supply infrastructure that is adequate today, if recharging takes place during off–peak periods, to power a significant proportion of the UK vehicle fleet. Its other prime advantage is that no vehicle exhaust emissions are generated at point–of–use.

Because present battery systems cannot compete in terms of range, and to a lesser extent performance, with petrol–fuelled vehicles, the future success of electric vehicles (EVs) depends on current battery research and engineering development. The outlook for significant improvements in commercial battery technology – especially regarding energy density and power, now appears more promising, given the huge collaborative efforts now directed at these issues. Nevertheless, there remain uncertainties about the costs and durability of advanced batteries, and previous predictions about imminent breakthroughs in battery technology have repeatedly proved incorrect.

Despite essentially zero vehicular emissions, EVs need to be assessed on the emissions impact from the power stations providing their electricity. Most power generation in the UK is currently from coal–fired stations, although the mix will change significantly over the coming decade as more natural gas–fired combined cycle gas turbine power stations come on line, reducing carbon dioxide and sulphur dioxide (SO_x) emissions especially and reducing the greenhouse impact for equivalent power generated. Although nuclear and hydroelectric power would be more desirable as recharging sources from an air quality point of view, they contribute a minority proportion to the UK national grid and the future of nuclear power is under considerable debate currently. Consequently, the use of EVs to replace conventionally–fuelled vehicles trades off a reduction in local (usually urban) air pollution against an increase in regional emissions and the long–range dispersion of NO_X and SO_X from the increased power generation.

Some recent EV designs, such as the General Motors "Impact" vehicle, may overcome some of the shortcomings generally associated with electric vehicles. The Impact achieves a substantial boost to its range by attaining very high levels of vehicle efficiency, incorporating an extremely aerodynamic shape (its drag coefficient, C_D, is 0.19 versus 0.30 for most modern passenger cars today) and low weight, low rolling resistance wheels and tyres amongst other measures. Achieving high vehicle efficiency is an important strategy for alternative fuels because of their lower energy density, and is particularly critical for EVs and hydrogen–fuelled vehicles.

12.9 Hydrogen

Hydrogen's primary appeal is its cleanliness – its vehicular use will generate no carbon dioxide (the principal greenhouse gas), very low emissions of hydrocarbons and particulates (from lubricating oil consumption), virtually no carbon monoxide or oxides of sulphur and moderate NO_X emissions. The main drawbacks are the high cost of the fuel, limited vehicle operating range (liquid hydrogen has one–sixth the energy density of petrol) and difficult and expensive on–board storage – either in heavy and bulky hydride systems that may adversely affect range and performance or in bulky cryogenic tanks that will reduce available cargo space.

In several ways, hydrogen vehicles share many pollution and performance characteristics with electric vehicles, but with the potential for rapid refuelling, countered by more difficult fuel handling. The development of improved vehicle efficiency technologies (low weight and aerodynamics, for example) are important for successful introduction of hydrogen–fuelled vehicles because of hydrogen's low energy density – as it is for electric vehicles. Presently, the least expensive source of large quantities of hydrogen (but still at substantially higher costs than petrol) is from fossil fuels, either from natural gas reforming or coal gasification (which would exacerbate greenhouse gas emissions).

Production of hydrogen from photovoltaic (PV) systems would yield an overall fuel supply system that generated virtually no greenhouse gases, but costs are still prohibitively high. Substantial cost reductions are required, such as those associated with improvements in PV module efficiency and longevity.

12.10 Conclusions

The conclusions that may be drawn from the review presented in this book are that conventional petroleum and diesel fuels, derived from crude oil, will continue to be used as the dominant motor fuels, at least for the next 20–30 years. It is probable that Europe will adopt certain fuel reformulations sometime in the future, although not necessarily in line with the present US proposals. Due to potential health concerns, the limitation of aromatic content (for petrol and diesel) and the reduction of benzene in petrol is likely, and octane is likely to be maintained by the addition of oxygenates (although the effects of oxygenates on emissions from European vehicles is largely unknown). Diesel will be "reformulated" with the adoption of a low sulphur content (≤0.05%) and may become more widely available with increased cetane and "sulphur–free" (<0.01%) for niche markets.

The prospects for conversion to alternative fuels are exerting pressure on the petroleum industry to devise petroleum–based solutions to many of the problems alternative fuels are expected to address. Although revisions to conventional (mostly petrol) fuel composition are unlikely to address the problem of fuel diversification and reduce dependence on imported oil, it appears to be air quality concerns that are driving the current interest in alternative fuels – and further changes to petrol can reduce air pollution. In the United States, ARCO's announcement in 1989 of a reformulated, pollution–reducing petrol as an alternative to regular leaded petrol in California was perceived as probably the first move in an industry effort to help draw attention and interest away from other alternatives.

Other alternative fuels are likely to find niche applications and gradually start to penetrate the market during the next 5–10 years, but not significantly at the expense of petrol and diesel fuels. Examples include urban bus and local delivery fleets operating on natural gas or LPG. Heavy–duty diesel engines, especially those used in buses, would benefit from very low particulate and low NO_x emissions from the use of alcohol fuels, helping reduce fears of the potential health problems from diesel particulates. Other alternative fuels, such as vegetable oils, bioethanol and other biomass fuels, are likely to find most application in developing countries and remote regions where conventional fuels are expensive or difficult to distribute. Rapeseed oil (and especially rape methyl ester) has found niches, especially in central European countries, as a useful product for absorbing surplus rapeseed production.

Electric vehicles (EVs) are likely to become more popular after the turn of the century, especially as the Californian EV (zero emission) market should be established and most vehicle manufacturers will offer EVs in order to preserve their conventionally–fuelled vehicle sales in that State (and possibly others). Hybrid vehicles may help to overcome consumers' reservations about using EVs, until battery and/or fuel cell development is at a stage where acceptable range and rapid recharging, accompanied by a reduction in their size and weight is achievable, and adequate infrastructure is in place. That is most likely to happen in 10 to 25 years time.

Hydrogen must be viewed as a long–term fuel (30–50 years hence) with potential emissions advantages, provided it can be manufactured cheaply and cleanly. Storage systems need considerable development effort to reduce their weight and size for a given energy storage. If significant electricity can be generated from renewable sources – tidal or hydro, wind, sunlight and biomass, the use of hydrogen as a gaseous vehicle fuel (in combustion engine or fuel cell) or the electricity stored in advanced batteries onboard an electric vehicle, appear to be the most likely future for motor fuels as fossil fuel reserves become increasingly exhausted later in the 21st century.

INDEX

ACEA   18, 19
ADEME   72, 75, 76
AGO   18
Alcohols
    ethanol   53-63
    methanol   29-47
Alfa Romeo   107
AQIRP   11, 13-16, 37
ARCO   10
Aromatics   10, 13-16, 192-194
Audi   131
Auto/Oil program, *see AQIRP*
AVL   58, 73
Avocet   35, 45, 57, 62

Batteries   124-130
    aluminium-air   129
    comparative performance   125
    iron-air   129
    lead-acid   126
    lithium aluminium-iron   128-129
    lithium solid polymer   129
    nickel-cadmium   126
    nickel-iron   127
    nickel-metal hydride   130
    sodium-nickel chloride   128
    sodium-sulphur   127-128
    zinc-air   129
    zinc-bromine   130
Benzene   10, 192-194
Biodiesel   67
Bioethanol   54, 55
Biofuels   67-79
BMW   131, 163
Brazil   33, 53, 54, 55, 62
British Gas   94, 111, 115
Butane, *see LPG*

California   33, 43, 44, 46, 116
Cetane   19, 20
Chrysler   44, 114, 116, 131
City Diesel   21
City Gasoline   21

Clean Air Act (US)   17
CleanAir Transport   131
Coal
    liquid fuels   22, 23
    reserves   6-7
    synthetic fuel   22-23
Compressed natural gas, *see natural gas*
Crude oil   4-5
    unconventional sources   22
Cummins   108, 109, 116

DAF   116
Daimler-Benz   159, 163
Detroit Diesel Corporation   38, 44
    58, 62, 116
Diesel fuel   18-21
    cetane   19, 20
    density   20-21
    sulphur   19-20
Dual-fuel   35, 42, 57, 61

EC-1   10
Electricity   123-150
    generation   138-144
    solar cells   143-144
Electric vehicles   123-150
    demonstration   135-138
    emissions   147-149
    energy consumption   144-147
    specifications   130-135
Emissions
    carbon dioxide   169-171
    exhaust   178-180
    fuel cycle   173-176
    greenhouse gases   171-173
    health effects   187-201
    production & distribution   177-178
EMPA   73
ETBE   11, 12, 15, 16
Ethanol   12, 14, 53-63, 190
    air toxics   59
    costs   60-61
    demonstration   62
    feedstocks   54-55

Ethanol .....
    fuel characteristics 53-54
    infrastructure 55-56
    life cycle emissions 59-60
    regulated emissions 57-59
    safety 54
    vehicle conversion 56-57
European Auto/Oil Programme 18

Fiat 131-132
Flexible fuel vehicle 15, 29, 34, 43
    44, 45, 54, 61, 62
Ford Motor Company 34, 44
    114, 116, 132
Fossil fuels
    reserves 3-7
    reserves/production ratio 3, 4, 5, 6
Fuel cells 139-143
    phosphoric acid 140
    proton exchange membrane 140-141
    solid oxide 142
Fuel costs 182-183
Fuel storage 180-182
Fumigation 35, 57

Gasohol 53
General Motors Corporation 44, 114
    116, 132
Greenergy 21

Health & environmental effects 187-201
    aldehydes 190-192
    aromatics 192-194
    benzene 192-194
    CO 195-196
    CO_2 200
    ethane 189-190
    ethanol 190
    HCs 187
    lead 198-199
    methane 189-190
    methanol 190
    NMHCs 187
    NMOGs 187-188
    NO_x 194-195
    olefins 192

Health & environmental effects .....
    ozone 199-200
    PAHs 192-194
    PAN 199-200
    particulates 196-198
    propane 189-190
    VOCs 188-189
Hybrid vehicles 123
Hydrogen 157-165
    costs 162-163
    demonstration 163-164
    emissions 161-162
    feedstocks 158-159
    fuel characteristics 157-158
    infrastructure 159
    safety 158
    vehicle conversion 159-161

ICI
    Avocet 35, 45, 57, 62
Institut du Pétrole 72
Iveco 109, 110, 133

Liquefied natural gas 100, 103, 115
Liquefied petroleum gas (LPG) 83-96
    air toxics 91-92
    costs 92-93
    demonstration 93-94
    feedstocks 85-86
    fuel characteristics 83-85
    infrastructure 86
    life cycle emissions 92
    regulated emissions 88-91
    safety 85
    vehicle conversion 86-88

MAN 38, 116
Mazda 163
Methane, *see natural gas*
Mercedes-Benz 45, 73, 77, 133
Methanol 12, 29-47, 190
    air toxics 40-41
    costs 42-43
    demonstration 43-45
    feedstocks 31-32
    fuel characteristics 29-30

Methanol .....
    infrastructure 32
    life cycle emissions 41-42
    regulated emissions 36-39
    safety 30-31
    vehicle conversion 32-36
Mitsubishi 133
MTBE 10, 11, 14-16, 117

Napthenes 10
Natural gas 99-117
    air toxics 110
    costs 111-113
    demonstration 113-115
    feedstocks 101
    fuel characteristics 99-101
    infrastructure 101-102
    life cycle emissions 110-111
    regulated emissions 105-110
    reserves 5-6
    safety 101
    vehicle conversion 102-105
Neste Oy 21
Nissan 45, 133, 136
Novamont 71, 77

Octane 12
Oil reserves 4-5
Olefins 10, 13, 14, 16, 192
Opel 90, 132
Oxygenates 11-13, 24

Paraffins 10
Petroleum 9-18
Peugeot-Citroën (PSA) 45, 72, 114
133, 137, 164
Phillips Petroleum 16-18
    Unleaded Plus 16
Photovoltaic cells 143
ProAlcool 62
Propane, *see LPG*

Rape methyl ester (RME) 67-79
Reading Buses 76, 77
Reformulated fuels 9-21
    commercial fuels 21

Reformulated fuels .....
    diesel 18-21
    oxygenates 11-13
    petrol 9-18
Renault 134
RVP 13, 14, 15, 16

Saab 45, 62, 134
Solar cells 143-144
Sulphur 10, 13, 14, 15, 16, 19-20
Synthetic liquid fuels 21-23

T_{10} 19
T_{90} 13, 14, 15, 16
TAME 11
TBA 11
TNO 34, 90, 91, 94
Toyota 134

UBA 74, 75
US Reformulated Gasoline Program 18
UTAC 72

Vapour pressure 10, 11, 12
13, 14, 15, 16
Variable fuel vehicle 15, 33, 61
Vegetable oils 67-79
    costs 75-76
    demonstration 76-77
    emissions 71-74
    feedstocks 69-70
    fuel characteristics 67-69
    infrastructure 70
    life cycle emissions 74-75
    safety 69
    vehicle conversion 70-71
Volkswagen 33, 39, 45, 134, 135
Volvo 45, 89, 114, 116, 135

Computational Mechanics Publications

Alternative Engines for Road Vehicles

M.L. POULTON, *Transport Research Laboratory, Crowthorne, UK*

Spark-ignition engines and compression-ignition diesel engines have been almost unchallenged as power units for road vehicles for the last century. However the air pollution problems which they cause, and their poor efficiency in today's typical operating conditions, make it essential that alternatives to these engines are fully explored. Many have been proposed, but few have been exploited commercially.

This book discusses the technologies that presently exist for alternative engines. It therefore provides a unique source of information for engineers, scientists and managers involved with vehicle development and planning. For each alternative engine that is considered, the operating principle is discussed, together with the primary advantages and disadvantages of the particular technology. The alternative engines and prospects for further development of conventional engines are dicussed and compared with reference to fuel economy and exhaust emissions.

This book forms part of a programme of study of the contribution of road transport to global warming and air pollution, and investigating ways of reducing it.

ISBN: 1853123005; 1562522248 (US, Canada, Mexico) May 1994 192pp £59.00/$89.00

Urban Air Pollution

Edited by **N. MOUSSIOPOULOS,** *Aristotle University of Thessaloniki, Greece,* **H. POWER** *and* **C.A. BREBBIA***, Wessex Institute of Technology, Southampton, UK*

This book covers mathematical modelling of mesoscale atmospheric dispersion.

Topics to be covered include: Urban Heat Island and Air Pollution; Photochemical Modelling of Pollution Scenarios in Mexico City; The Reactivity of Emissions, the Set of Conditions prevailing in Mexico City; Intense Radiation; Spatial and Temporal Evaluation of Automotive Emissions for an Urban Area; T Emissions on the Mesoscale Environment; Air Pollution in Athens; Landfill Emissions of Gases to the Atmosphere; Importance of Transport Processes in the Urban Air Quality; Air Pollution Levels and Trends from 1972 to 1992 in Belgium; Atmospheric Pollution in the Lisbon Airshed; Air Quality Study for the City of Graz, Austria.

ISBN: 1853123315; 1562522558 (US, Canada, Mexico) Oct 1994 apx 300pp apx £79.00/$118.00

Computer Techniques in Environmental Studies V

Edited by: **P. ZANNETTI***, Failure Analysis Associates Inc., Menlo Park, California, USA*

This book contains the proceedings of the Fifth International Conference on Development and Application of Computer Techniques to Environmental Studies, being held in the USA in November 1994. Topics to be covered include: Pollution: Air, Water, Soil, Noise, Radiation; Mathematical Modelling; Environmental Sciences and Engineering; Chemistry, Physics and Biology, Meteorology and Climatology; Fluid Dynamics; Remote Sensing; Software Implementation.

ISBN: 1853122726; 1562521969 (US, Canada, Mexico) Nov 1994 apx 700pp apx £148.00/$220.00

Environmental Modeling

Volume 2

COMPUTER METHODS AND SOFTWARE FOR SIMULATING ENVIRONMENTAL POLLUTION AND ITS ADVERSE EFFECTS

Edited by **P. ZANNETTI***, Failure Analysis Associates, California, USA*

This title is the second volume in an edited series of publications on computer methods for simulating environmental pollution and its adverse effects. It presents a careful selection of invited review papers covering environmental modeling topics. Each chapter has been authored by a leading scientist in the field and provides the reader with an organized and consistent discussion on the field of mathematical and numerical simulation of environmental phenomena. In addition, 'Environmental Modeling, Volume II' provides a critical review of software available for environmental simulations.

Partial Contents: Introduction to Environmental Modeling; Aerial Spray Drift Modeling; Survey of Long Range Transport Models; Finite Element Modeling of the Transport of Reactive Contaminants in Variably Saturated Soils with LEA and non-LEA sorption.

Series: Environmental Modeling, Volume 2

ISBN: 1853122815; 1562522051 (US, Canada, Mexico) March 1994 384pp £98.00/$148.00

All prices correct at time of going to press. All books are available from your bookseller or in case of difficulty direct from the Publisher.

Computational Mechanics Publications
Ashurst Lodge, Ashurst, Southampton,
SO40 7AA, UK.
Tel: 44 (0)1703 293223 Fax: 44 (0) 1703 292853

Computational Mechanics Publications

Computers in Railways IV

Edited by: **T.K.S. MURTHY,** *Wessex Institute of Technology, UK;* **B. MELLITT,** *London Underground, UK;* **C.A. BREBBIA,** *Wessex Institute of Technology, UK;* **G. SCIUTTO,** *Universita degli Studi di Genova, Italy and* **S. SONE,** *University of Tokyo, Japan*

These two volumes contain the edited proceedings from the Fourth International Conference on Computer Aided Design, Manufacture and Operation in the Railway and other Mass Transit Systems held in September 1994.
SET ISBN: **T.1853122661; 156252190X (US, Canada, Mexico) 1112pp £255.00/$379.00**

Railway Design and Management
Volume 1

Covers railway planning, management and information systems, design, manufacturing and testing as well as computer modelling and simulation of systems.
ISBN: **1853123544; 1562522825 (US, Canada, Mexico) August 1994 600pp £151.00/$2226.00**

Railway Operations
Volume 2

Covers railway operations, signalling communications and advanced train controls, traction and power supply as well as high speed rail and Maglev systems.
ISBN: **1853123595; 1562522833 (US, Canada, Mexico) August 1994 512pp £129.00/$193.00**

Air Pollution II

Edited by **J.M. BALDASANO,** *Universitat Politecnica de Catalunya, Spain;* **C.A. BREBBIA,** *Wessex Institute of Technology, UK;* **H. POWER,** *Wessex Institute of Technology, UK;* **P. ZANNETTI,** *Failure Analysis Associates Inc., USA*

The last decade has shown an increase in public and government concern around environmental issues due to air pollution, in particular those generated by man-made processes seeking the comfort of modern society. Atmospheric pollution consists of the adverse effects on the environment of a variety of substances (contaminants) emitted into the atmosphere by natural and man-made processes. It is a multifaceted problem that includes phenomena involving the different scales; a near-field phenomena that governs the way in which the contaminant rises, short-range transport in which the ground-effect is predominant, an intermediate transport where chemical reactions become important and a long-range transport where the decay and deposition effects are relevant. These books contain the papers presented at the Second International Air Pollution Conference held in September 1994.
ISBN: **1853122718; 1562521950 (US, Canada, Mexico) September 1994 1184pp £248.00/$372.00**

Computer Simulation
Volume 1

Partial Contents: Meteorological Modelling; Turbulence and Diffusion Modelling; Chemical Transformation; Urban Case Studies; Pollution Engineering.
ISBN: **1853123609; 1562522841 (US, Canada, Mexico) September 1994 608pp £140.00/$210.00**

Pollution Control and Monitoring
Volume 2

Partial Contents: Global Studies; Emission Inventory and Modelling; Data Analysis and Observation; Monitoring and Laboratory Studies; Pollution Control and Management.
ISBN: **1853123617; 156252285X (US, Canada, Mexico) September 1994 576pp £133.00/$199.00**

Air Pollution

Edited by **P. ZANNETTI,** *Failure Analysis Associates Inc, California, USA,* **C.A. BREBBIA,** *Wessex Institute of Technology, Southampton, UK,* **J.E. GARCIA GARDEA,** *Instituto Tecnologico y de Estudios Superiores de Monterrey, Mexico and* **G. AYALA MILIAN,** *Universidad Nacional Autonoma de Mexico, Mexico*

This book covers the latest research information on monitoring, simulation and management of air pollution problems. It contains papers presented at the Air Pollution Conference held in 1993.
Contents: Numerical Modelling; Meteorological Modelling; Turbulence and Diffusion Modelling; Chemical Transformation Modelling; Global Studies; Analysis, Monitoring, Management and Engineering; Data Analysis and Observation; Monitoring and Laboratory Studies; Pollution Management; Pollution Engineering.
ISBN: **185312222X; 1562521462 (US, Canada, Mexico) February 1993 808pp £200.00/$360.00**

Air Pollution Modeling
THEORIES, COMPUTATIONAL METHODS AND AVAILABLE SOFTWARE
P. ZANNETTI, *AeroVironment Inc., Monrovia, USA*
ISBN: **1853121002; 094582484X (US, Canada, Mexico) November 1990 448pp £64.00/$118.00**

All prices correct at time of going to press. All books are available from your bookseller or in case of difficulty direct from the Publisher.

Computational Mechanics Publications
Ashurst Lodge, Ashurst, Southampton,
SO40 7AA, UK.
Tel: 44 (0)1703 293223 Fax: 44 (0) 1703 292853